WINDOWS ON MATHEMATICAL MEANINGS:
LEARNING CULTURES AND COMPUTERS

Mathematics Education Library

VOLUME 17

The titles published in this series are listed at the end of this volume.

WINDOWS ON MATHEMATICAL MEANINGS
LEARNING CULTURES AND COMPUTERS

by

RICHARD NOSS & CELIA HOYLES
University of London, U.K.

KLUWER ACADEMIC PUBLISHERS
DORDRECHT / BOSTON / LONDON

Library of Congress Cataloging-in-Publication Data

```
Noss, Richard.
   Windows on mathematical meanings : learning cultures and computers
/ by Richard Noss & Celia Hoyles.
      p.   cm. -- (Mathematics education library ; v. 17)
   Includes index.
   ISBN 0-7923-4073-6 (hb : acid free papers)
   1. Mathematics--Study and teaching.   I. Hoyles, Celia, 1946-
II. Title.   III. Series.
QA11.N65   1996
510'.71--dc20                                          96-17822
```

ISBN 0-7923-4073-6

Published by Kluwer Academic Publishers,
P.O. Box 17, 3300 AA Dordrecht, The Netherlands.

Kluwer Academic Publishers incorporates
the publishing programmes of
D. Reidel, Martinus Nijhoff, Dr W. Junk and MTP Press.

Sold and distributed in the U.S.A. and Canada
by Kluwer Academic Publishers,
101 Philip Drive, Norwell, MA 02061, U.S.A.

In all other countries, sold and distributed
by Kluwer Academic Publishers Group,
P.O. Box 332, 3300 AH Dordrecht, The Netherlands.

Printed on acid-free paper

Printed in the Netherlands

CONTENTS

FOREWORD

This book is the culmination of some ten years' theoretical and empirical investigation. Throughout this period, we have come into contact with many who have stimulated our thinking, some of whom belong to the community of Mathematics Educators. Our membership of that community has challenged us to make sense of some deep issues related to mathematical learning, especially the cognitive and pedagogical faces of mathematical meaning-making.

Alongside this community, we are privileged to have been part of another, whose members are centrally concerned both with mathematics and education. Yet many of them might reject the label of Mathematics Educators. This community has historically been clustered around what is now called the Epistemology and Learning Group at the Massachusetts Institute of Technology. Their work has focused our attention on cognitive science, ethnography, sociology, artificial intelligence and other related disciplines. Crucially, it has forced our awareness of the construction of computational settings as a crucial component of the struggle to understand how mathematical learning happens.

We have sometimes felt that few have tried to span both communities. Indeed, an analysis of the references in the literature would, we are sure, reveal that the two communities have often ignored each other's strengths. One reason for writing this book is born of our hope that we might draw together Mathematics Educators and mathematics educators, and assist both communities in recognising that there are insights that might be derived from each other.

One group of scholars who have consistently shared this dual aim has been the Logo and Mathematics Education group, founded in London in 1985, and a constant source of inspiration and insight. To them, our thanks.

We owe our greatest intellectual debt to Seymour Papert, whose ideas have provided continual inspiration, and whose support and friendship has been invaluable throughout the last fifteen years.

Others who have, at various times, formed part of the MIT community, have similarly informed our approach, and generously – sometimes unknowingly – helped us towards whatever pretence to theoretical clarity we now claim. Prime among these are Andy diSessa, Brian Harvey, Edith Ackermann, Sylvia Weir and Uri Wilensky. We especially thank Sylvia Weir for first drawing our attention, more than a decade ago, to the power of the window metaphor. Not far from MIT, either geographically or intellectually, others who have provided us with a rich vein of ideas include Al Cuoco, Paul Goldenberg, Wally Feurzeig and many others at TERC, EDC and BBN. In these communities, and especially in the work of Cuoco, Goldenberg, Papert and Wilensky, we have found a sense of intellectual resonance which has helped to clarify many of our own ideas.

We thank the following for their help and invaluable suggestions for improvement on an earlier draft of the manuscript: Al Cuoco, Sandy Dawson, Brian Harvey, Lulu Healy, Colette Laborde, Seymour Papert, and Stefano Pozzi. Our thanks too, to an anonymous reviewer who pointed out the need for some revisions, and whom we hope will now come out of the closet so that we can thank him or her personally.

We would like to thank all our colleagues and students in the Mathematical Sciences Group at the Institute of Education, University of London, for helping to create a scholarly, supportive, and collaborative milieu. We must single out for special thanks, Lulu Healy and Stefano Pozzi, for their help with some of the studies referred to in the book, and for their generosity in sharing their insights with us over many years. Thanks too, to Rosamund Sutherland, for her collaboration in the work on the *Microworlds* project.

We have worked individually and jointly on a large number of projects over many years, and part of the corpus of data we have collected has found a place in this book. It is our sense that the theoretical fruit of this personal and intellectual collaboration has been greater than the sum of its constituent parts.

Celia Hoyles, Richard Noss
London
March 1996

ABOUT THE AUTHORS

Richard Noss studied mathematics at the University of Sussex, England, receiving his Bachelor's degree in 1971 and his M. Phil (in Geometric Topology) two years later. He taught in London secondary schools until 1982 when he began researching programming-based mathematical environments for children. He received his Ph.D. in 1985, and in the same year was appointed to the Institute of Education, London University. He has remained there ever since, and is now Professor of Mathematics Education.

Celia Hoyles was awarded a first class honours degree in mathematics in 1967 from the University of Manchester, England after which she taught in London schools and studied for her M.Ed in mathematics education. She moved to the Polytechnic of North London in 1972 and obtained her Ph.D. in 1980. In the early 1980's, she started to study the use of computers in mathematics teaching and learning. This has remained a focus for much of her research. She became Professor of Mathematics Education at the Institute of Education, University of London in 1984.

CHAPTER 1

VISIONS OF THE MATHEMATICAL

In North Greenland distances are measured in *sinik,* in 'sleeps', the number of nights that a journey requires. It's not a fixed distance. Depending on the weather and the time of year, the number of *sinik* can vary. It's not a measurement of time either. Under the threat of a storm, I've travelled with my mother non-stop from Force Bay to Iita, a distance that should have required two nights.

Sinik is not a distance, not a number of days or hours. It is both a spatial and a temporal phenomenon, a concept of space-time, it describes the union of space and motion and time that is taken for granted by the Inuit but that cannot be captured by any European everyday language.

The European measurement of distance, the standard meter in Paris, is something quite different. It's a concept for reshapers, for those whose primary view of the world is that it must be transformed. Engineers, military strategists, prophets. And mapmakers. Like me.

Miss Smilla's Feeling for Snow (Høeg, 1994, p. 278)

Mathematics: the science of space and number, the study of pattern and structure, the queen of sciences. To think mathematically affords a powerful means to understand and control one's social and physical reality. Yet despite some twelve or so years of compulsory mathematical education, most children in the developed world leave school with only limited access to mathematical ideas, or much affinity with the idea of taking a mathematical point of view. In many cultures including our own, it is quite acceptable – even fashionable – to admit ignorance of things mathematical in ways which would be inexcusable in relation to art, literature or music.

From the individual's point of view, alienation from mathematical ways of thinking may be regrettable, though hardly terminal. It might lead to a one-sided appreciation of social and physical phenomena, but there are other, equally engaging sides to be accessed. It might limit opportunities for employment, though this is often exaggerated. It will close down any deep appreciation of things scientific, but the illiterate scientist has as much to lose in terms of his or her humanity than the innumerate artist.

No. The real impoverishment is cultural not individual. For the social and material world is becoming increasingly mathematised, a world in which – in Weizenbaum's (1984) memorable phrase – judgement is increasingly replaced by calculation. It is often argued that increasing mathematisation leads to a rising profile of mathematical ideas within the culture. In fact, the reverse is true. Mathematics is becoming all-pervasive: but it is mathematics which lies deep inside the machines which govern our existence, invisible mathematics which lies dormant in the electronic circuits of our washing machines, buried, for example, within the communication networks which so powerfully structure our view of the world. It is mathematics produced by the few, designed for the consumption – perhaps the control – of the many. From this imbalanced state emerges a danger that we, as individuals and communities, are losing our ability, even our will, to understand the systems in which we live. In the extreme, if the human race comes to an abrupt end, it is certain that collective judgement will have been suspended in favour of the iron laws of calculation embodied in computers. The realm of social and cultural life, as much as the quality of life of the individual, is impoverished by the absence of mathematical windows to understand and interpret the world, and to change it.[1]

Paradoxically, the computer will form a major theme in what follows. It will afford us insights into people's mathematical meanings, and simultaneously allow us to glimpse new mathematical epistemologies which will facilitate their construction. It will provide a setting to help us make sense of the psychology of mathematical learning, and at the same time, afford us an appreciation of mathematical meaning-making which takes us well beyond the psychological realm. Above all, the computer will open new windows on the construction of meanings forged at the intersection of learners' activities, teachers' practices, and the permeable boundaries of mathematical knowledge.

There is, as we have said, a paradox. It is far from obvious how the computer – the epitome of dehumanised social relations in our culture – could aid in the process of making mathematics more accessible, more learnable. After all, the construction of meaning is a human quality – *the* human quality – and to make mathematics more meaningful is simultaneously to humanise it. We hope that by the end of this book, we will have clarified just how our vision makes room for the computer – at least the computer conceived in rather special ways – to take on this role.[2]

Our concerns in this book range over the epistemological, cognitive, cultural and methodological. In the remainder of this chapter, we will introduce some of these organising themes, and will outline our rationale for observing the construction of mathematical meanings through the window of the computer.

1. RESHAPING MATHEMATICS, REVISIONING LEARNING

Miss Smilla's distinction is familiar enough. There are two kinds of knowledge mediated by language. One is designed for action, for practices which are embedded in the culture. The other is used to describe concepts appropriate for 'reshapers', those who wish not merely to act in the world, but to transform it.

This distinction, as we shall see in Chapter 2, has structured much of the thinking about mathematics education; it is premised on epistemological assumptions, but we shall see that it readily spills over into beliefs concerning the nature of learning and teaching. We, like Smilla's engineers and prophets, believe in the power of the reshapers' perspective; but we believe too that it is possible to achieve this power without rejecting the ideas and language of cultural practices. If we are right, we will have indicated not only a way to bridge Smilla's epistemological divide, but to point to the possibility of opening access to the reshapers' view of the world to more than the few.

In order to clarify how people think with, and about, mathematics, we have to find some way of recognising mathematical activity when we encounter it. One way would be to provide a neat definition of what mathematics, or school mathematics (they are not the same) actually is. But in reality this would be less than helpful, as it would prejudge some critical issues which we intend to explore. Is there a unitary mathematics? Is there only one way of understanding the world mathematically, or does it make more sense to blur the boundaries of what constitutes mathematics? If the latter, how far can we go before we subsume under the heading 'mathematical' all intellectual endeavour? In order to answer these and similar questions, we will describe activities which count, for us, as mathematical, scrutinise these for the meanings that they evoke and thus delineate the sometimes fuzzy boundaries of mathematics. When we think of mathematics, we think of it as an activity; and when we ask the question 'what is mathematics?', we intend the question not merely in the philosophical sense, but in the sense of what it means to *do* mathematics, and just as importantly, how it might be done.

This dual perspective will involve us in a quest which is at once psychological and epistemological. Accordingly, while we offer visions of mathematical learning in which we will try to access the cognitive aspects of mathematical meaning-making, we also seek to *re*-vision mathematical learning, to suggest ways we might construct new, mathematically rich experiences for learners. The notion of revisioning encompasses more than re-examination or reconsideration – for these suggest that what we are revising is a given, a fixed goal whose locus of interest revolves around the approach taken to achieve it. We would like to scrutinise the goal itself – the vision – of what mathematics education might be; and we would like to explore how these revisions might shape and be shaped by the strategies and approaches which are used to achieve it.

There are numerous ways in which school mathematics has been reconceived in the recent past including 'new' mathematics, 'problem solving', 'investigations', 'real' maths; to varying degrees, these have offered various panaceas of theory and practice which, it has been argued, would solve the fundamental dilemmas confronting mathematical learning and teaching. For the most part, such attempts have centred attention on pedagogic matters; epistemology has been implicitly or explicitly ignored in favour of 'solutions' to 'problems': pedagogy without epistemology. Where there has been an epistemological dimension, it has tended to base itself outside school mathematics, in, for example, an analysis of the mathematical needs of society or the mathematical requirements of the university. Thus, efforts to fashion new school curricula have tended to import ready-made visions of mathematics from elsewhere, presenting a new unattainable mountain to scale rather than an authentic attempt at creating an alternative epistemology.

There has, however, been a radical approach which has taken students and their relationship with school mathematics as the starting point of curriculum development. This strand is represented by the early constructivist movement, by the pioneers of investigative mathematics, and by those at the leading edge of work concerned with the potential of computers in mathematical settings, in and around the Logo community at the Massachusetts Institute of Technology. An early slogan of this latter community was:

'Let's stop trying to make kids learn math; let's make mathematics for kids to learn.'

By stressing the epistemological dimension, it focuses attention on the need to create, to construct experiences which are both appropriable by students *and* genuinely mathematical. The slogan forces us to accept a broadening of what counts as mathematical and to democratise the often implicit assumption that mathematics is attainable by only the chosen few. Implicit in the formulation is that it is 'traditional' school mathematics that we must stop trying to teach, and that we must divert some of our energy towards building new and diverse kinds of mathematics which are at once learnable and accessible.

In this interweaving of epistemological and psycho-pedagogical concerns, we will only rarely be interested in the state of the mathematical art as practised by mathematicians, and we will often mean by 'mathematics', some version of what is – or might – be taught to and learned by young people. That young people in 'Western' societies spend much of their waking life in schools, will imply that our focus is often on learning and teaching in a professional sense; but this should not be blind us to the undoubted fact that much, perhaps most, learning takes place outside the school, and that schools as they currently exist are only a recent organisational structure which may not endure for ever in their current form.

2. COMPUTERS...

Our central heuristic is to take the idea of the computer as a window on knowledge, on the conceptions, beliefs and attitudes of learners, teachers and others involved in the meaning-making process. Yet our relationship with the computer is ambivalent in an important sense. In one respect, the computer will loom large in what follows; all of our own work which we discuss in this book involves the computer in one guise or another. Yet in another sense, we push the computer into the background. Our focus of interest is not the computer, it is on what the computer makes possible for mathematical meaning-making.

By offering a screen on which we and our students can express our aspirations and ideas, the computer can help to make explicit that which is implicit, it can draw attention to that which is often left unnoticed. At a basic level, this is a question of cultural assimilation rather than accommodation: 'Computers can make the things we already do badly in schools even worse, just as they can make many things we already do well infinitely better' (Smith, 1986, p. 206). This is true, but it stresses quantitative rather than qualitative change. Much more fundamental change is possible; the computer, as we shall see, not only affords us a particularly sharp picture of mathematical meaning-making; it can also shape and remould the mathematical knowledge and activity on view.

We will dwell briefly on this mediating role of the computer (or rather, on the mediating role of the linguistic and notational frameworks it provides). At this early stage, we can afford to be simplistic. Let us schematically pose the basic problem of learning and teaching as one of helping a learner to elaborate, extend and modify what is known. Whether we take a naive transmittal view of learning, or a more constructivist perspective, the difficulty is this: how to help B to know what A knows.

A —————————▶ B

Figure 1.1. A wants B to know what A knows. Ideally, B wants to know what A knows.

Now suppose a computer is introduced into this setting. Not just any computer and not just any software: let us for the moment assume that we have a rather special kind of computer, one on which B can express what he or she knows (we will elaborate what we mean by this in Chapter 3). Already A has some insight into the state of B's understandings, the meanings which are evoked for B by the problem or the situation. What is more, A now has a language with which to interact with B, the language of action in which ideas on the computer are expressed. Now let us further imagine that there is a culture, a setting in which A and B view learning as a mutual process in which pieces of mathematical knowledge can be co-constructed. The computer can act in a mediating role, shaping and moulding what A and

B know, perhaps even towards some point of convergence. A and B both have a two-way channel of communication with the computer, and in establishing these channels, it (actually the setting) has opened a channel from B to A where previously the direction of communication was essentially one-way.

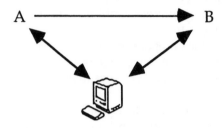

Figure 1.2. The computer acts in a mediating role, by providing a medium for linguistic and notational expression.

Now we are not so naive as to suggest that the computer is necessary for learning, or that it is unique in its ability to catalyse mathematical communication. We know perfectly well that sensitive teachers have always tried to open two-way lines of communication, and we recognise that they have often done so without computers.[3] Is there anything special that the computer has brought to the situation? Not the imparting of 'information'[4] from machine to person, but an enhanced communication between people; not the transmission of A's understanding to B, but an arena in which A and B's understandings can be externalised; not a means of displaying A's knowledge for B to see, but a setting in which the emerging knowledge of both can be expressed, changed and explored. But even this is not fundamental. The central difference is one of expression. For the language that A and B can now use to communicate is the language of the medium, the language of the computational system. In the non-computer setting, A and B would have had a choice between natural language or mathematical language: the former good for communication in general, but badly tuned to rigorous and precise discourses like mathematics: the latter precisely the reverse. The computer affords a half-world in which effective communication and precision[5] can be approached, where articulation and rigour can be made to converge. We shall have much more to say about this crucial issue in later chapters, and especially in Chapter 5.

The question of articulation[6] is key. For the computer's potential to reshape the learner-teacher relationship is only part of the picture. The fact that the computer demands a semi-formal articulation of relationships, implies that the computer offers a window, not just onto the teacher and learner, but onto the observer as well. It affords a view of the meaning-making process, a glimpse of learning which is often difficult, if not impossible, to catch. The computer provides a screen on which learners can express their thinking, and simultaneously offers us the chance to glimpse the traces of their thought.

The computer's presence opens a further, epistemological window on mathematical learning. It confronts us with fundamental questions about the nature of mathematical knowledge. For example, what is algebraic about Logo programming; what is geometric about dynamic geometry systems such as *Cabri Geometry?*[7] Is CabriGeometry a flavour of geometry, or is it something else? Is LogoAlgebra a pathway to traditional algebra, or a substitute for it? And if the latter, what have we lost or gained by this substitution?

The epistemological dimension has an opposite face. If we are challenged to consider the nature of computer-based mathematical knowledge, we will inevitably be drawn to consider new domains of mathematical activity built with the help of the computer. The task of understanding the meanings of computational mathematical settings, is simultaneously a challenge to create new mathematical scenarios based on what the computer can bring.

3. ... AND CULTURES

Our task is complex: we would like to steer a course between furthering our understanding of mathematical learning, and revisioning what mathematical learning comprises. The key to this complexity is to take seriously the notion of culture. In fact, we shall see in the following chapter, that ignoring the socio-cultural dimension leads us rapidly into an analytical cul-de-sac. We might as well recognise at the outset that we cannot analyse all the inter-relationships between learners, teachers, and mathematical knowledge: at best, we can try to understand the complexity of lived-in cultures from what we see, while keeping in mind Johnson's (1993) warning that 'It is tempting, but completely illegitimate, to infer lived effects from structural analysis' (p. 258).

Our focus on culture stems from two sources. In the first place, we will need to problematise the computer's presence in terms of tools and their relationship to the development of meaning. This will inevitably involve us in an investigation of broader cultural concerns, and in particular, to consider work which has conceptualised language as an intellectual tool. This socio-cultural perspective will allow us to consider mathematical meaning-making as more than a diad consisting of learner and knowledge; we will have much to say on the mediating influence of the tools at hand.

Our second source arises from a recognition of the diversity of learning experiences which abound in the literature. This issue came recently to our attention as we read through the *Handbook of Research on Mathematics Teaching and Learning*, edited by Doug Grouws (1992). This is a huge work commissioned by the National Council of Teachers of Mathematics in the US. In his introduction, Grouws makes the following point:

Research on mathematics teaching and learning has flourished over the past two decades. There now exists a recognisable body of research that not only is conducted within the realm of mathematics but also takes the nature of the mathematics domain into account in all aspects of the work, including framing research questions, choosing a mode of investigation, designing instruments, collecting data, interpreting results, and suggesting implications.

(Grouws, 1992, p. ix)

We set to thinking what, if anything, we could bring that was new: after all, Grouw's list is comprehensive, including the nature of mathematics, methodologies, and implications for practice. Yet as we read the book we suffered from recognition failure: we became aware that there was much in the book that was quite different from our own experience, despite sharing many of its general concerns and predilections. Quite simply, we felt we were in an alien culture. On reflection, this is, perhaps, not surprising. The book is unashamedly written from a United States perspective: 22 out of the 29 chapters are written by authors from the US, and the overwhelming majority of references to research are from the US.

We have no complaint with this focus, especially as a substantial body of high-quality research in mathematics education emanates from the USA. Our point is not that this vision of the educational world is American: it is that it is easy to mistake the implicit vision of American educational culture as educational culture *per se*. We started wondering about the heterogeneity of classroom cultures, of what is meant by 'math', and how learners are perceived – a heterogeneity and diversity which is evident between different classrooms in a single school, let alone across different continents. As we shall see, this diversity is made more evident and, perhaps, more understandable, when viewed through the window of the computer.

We drew the lesson that we cannot realistically talk of mathematical knowledge, mathematical meanings, or mathematical pedagogies *in general*. And we learned, more generally, that if we are to understand the process of mathematical meaning-making, we will need to study interactions between learners, between teachers and mathematical knowledge, and among players in the educational setting. Even then, we know that we are not able to do justice to the enormity of the task.[8] What we can do is to broaden our focus beyond the individual learner, and this we do in a variety of ways. We consider mathematical learning in collaborative settings (Chapter 6), how curricular change and broader educational and scholarly world-views influence each other (Chapter 7), the reciprocal influences of teachers' cultures on their attitudes and beliefs (Chapter 8), and the multiple intersections of diverse cultures in school (Chapter 9).

4. WINDOWS ON METHODOLOGIES

We have outlined a complex mix of cognition, culture and computers, and we must briefly outline the final, methodological, theme which allows us to make sense of this complexity.

The computer is a new technology: its very newness, together with the conceptual tools it makes available, forces attention on the structures and relations underlying the settings in which it is used. This has encouraged us to adapt old methodologies and to search for new ones. In part, this is simply a recognition that we are only now beginning to understand what questions to ask of computationally-mediated settings, let alone how to set about optimally answering them. But it also stems from our dual concerns to understand and construct, to gain insight into mathematical learning, and simultaneously to devise new settings for learning to take place. The methodological corollary is that we have sought to describe what people learn as well as how they learn it, in ways that respect the integrity of each goal. This has taken us along paths which, for example, involve reconceptualising the relationship between qualitative and quantitative research techniques, and looking for analytical constructs which point beyond computational settings altogether.

In general, our dual focus centres on studying change (to the extent that Chapters 7, 8 and 9 deal explicitly with curricular innovation in one form or another). In order to explain what we mean, we will give a mathematical analogy. Think of the positive integers {1, 2, 3, 4 ...}. They are familiar, they are known by us, we feel they are old friends. This knowledge does not come from staring at the familiar shapes of the numbers; neither does it stem from mere counting; it emerges from *acting on* the integers. We know the integers because we know how to add, subtract and multiply them. We have a sense of their size because we can compare them. And we know, for example, that some (mental) actions on the integers take us outside the realm of the integers altogether. We study the relationships *between* the integers, not the integers themselves. We know the integers because we can see what happens when we set them in motion.[9]

And so it is with studying someone else's thinking. We might like to be endowed with some special mental apparatus which would give us a representation of another's mental state, but such is not available. Neither can we hope to take a mental snapshot of what is 'known' at a point in time (although much 'testing' is undertaken in the vain hope that this is either possible or useful, or both). Instead we can set thinking in motion, and try to study what happens; we can set ideas in turbulence and investigate how changes occur; we can introduce new notions and try to understand how the thinker connects these with what he or she already knows. Within educational discourse, the study of thinking tends to presuppose that what is to be learned is fixed; the study of thinking-in-change demands that we devote at least equal attention to what is to be learned, as well as the meanings the learner draws from the educational experience.

The computer offers us a particularly pertinent means to adopt this approach: it is perhaps not an exaggeration to say that it is the most persuasive argument for our focus in this book. Throughout the following chapters, we have interspersed descriptions of our own work (almost entirely exploiting the computer's presence), with that of other workers in the field of mathematics education (both with and without the computer). We hope this will achieve two goals. First, we aim to construct a theoretical framework in which we can make sense of mathematical meaning-making, and we wish to acknowledge the intellectual debt we owe to researchers in the field. It will also allow us to locate our own position within the various research communities.

Alongside this straightforward aim is another, more subtle one. We are aware of an implicit barrier between the study of mathematical learning in computer-mediated settings, and that which employs traditional, inert media. We are acutely aware of how much our own understandings have been shaped by those whose work falls into the latter category; but we are equally aware that some important work that falls within the 'computational' framework, has not so far penetrated the broader concerns of the mathematics education community.

There may be many reasons for this. One, of course, might be the poor quality of the research output. We think this a little unfair, and we will refer to many in the 'computer-maths community' who have provided important insights which point beyond the computer. It might be that the concerns of the two communities are somewhat different: and for some part of the last two decades, this might well have been the case. Or there might be a deeper reason; it may be that with any new technology (film, video, or 'hyper-media'), initial attention is focused on the technology itself, rather than on what might be done with it. Even those who resist this temptation can easily be ignored as sharing technical, rather than educational, goals; and perhaps they (we?) have not always been guiltless in this respect.

It is for this reason, if for no other, that the window metaphor is important. Windows are for looking through, not looking at. It is true that windows mediate what we see and how we see it. Equally, windows can, at times, be objects for design and study. But in the end, what counts is whether we can see clearly beyond the window itself onto the view beyond.

NOTES

1. This is an appropriate point to acknowledge the profound influence of the notion of 'window' which we first encountered in Sylvia Weir's work, and which is discussed in her book *Cultivating Minds* (Weir, 1987).
2. This paragraph epitomises the difficulty of the term 'the computer'. We no more believe that 'the computer' will allow us to understand mathematical learning,

than we believe that 'the pencil' will allow us to write a better book. Yet finding some linguistic formulation which makes this point would achieve clarity at the expense of clumsiness. We will point to this problem once or twice in what follows, but in the end, our success in resolving it will have to be judged by the content of our argument, rather than the use of some alternative term.

3. It might also be pertinent to add that we have seen hundreds of cases where the computer's major contribution has been the closing down of communication altogether. Here again, is the danger of referring to 'the computer'.

4. As we might erroneously expect from the misunderstood appellation of 'Information' Technology (see Roszak, 1986).

5. We leave this idea undefined, together with associated ideas like *rigour* and *formalism*, which we will discuss in due course.

6. By articulation, we are being deliberately loose, stressing the linguistic, symbolic, and notational facets of the computer's role. A central part of the work of later chapters will be to tighten up this notion.

7. We will adopt Cabri Geometry (Baulac, Bellemain, & Laborde, 1994) as a generic dynamic geometry tool, as it happens to be the one we have used in our work. Other similar systems are The Geometer's Sketchpad (Jackiw, 1990), and The Geometry Inventor (Brock, Cappo, Arbel, Rosin, & Dromi, 1992).

8. For example, we do not, in this book, compare mathematical learning across cultures; neither do we give any serious attention to the political concerns which have shaped UK mathematical learning so greatly over the past decade.

9. This analogy is deliberately couched in terms which do not mention abstract algebra. On the other hand, we note that abstract algebra was invented precisely to exploit the power of describing transformations and 'movement'.

CHAPTER 2

LAYING THE FOUNDATIONS

1. INTRODUCTION

In this chapter we begin to focus on a central question of this book: how mathematics acquires meanings and how these are generated from various perspectives arising from different lived-in cultures. These meanings are multiply-derived and correspondingly diverse: yet in their diversity as well as in their commonalities, we hope to throw light on the fundamental issues raised in the previous chapter, and to open a window on what it is to mean mathematically. Our project will be to problematise some of the standard categories and dichotomies which have characterised thinking in the field of mathematics education: our methodology will involve charting a course between a range of empirical findings (our own and others') and the corpus of literature which has evolved within the academic community. We will lay the foundation for understanding the construction of mathematical meanings, an understanding which we hope to deepen as the book unfolds.

Focusing on meanings has three consequences. First, it forces us to think more clearly about what mathematics is. After all, if we expect to be able to recognise mathematical expression, we had better be clear how we would know it if we saw it. Just what counts as mathematics is, of course, a subject of debate among philosophers and (to a lesser extent) among mathematicians. And although seldom made explicit, many of the issues which underpin discussion of school mathematics are the same as those which underpin academic mathematics. Second, addressing the question of meaning opens up the possibility that mathematics as well as its educational counterpart has blurred boundaries; that both are constantly in (different) processes of change, that new mathematical ideas can emerge which do not conform to traditional norms (as, for example, specified in school or university curricula), yet which might be valuable either in their own right or as precursors to future learning. Third, it encourages us to look at the ways in which children express mathematics rather than to focus exclusively on what mathematics is learned. While the latter question continues to receive plenty of attention, our claim is that a focus on the delivery of knowledge overshadows the construction of meaning, and clouds rather than clarifies any understanding of how such meaning is constructed.

From a philosophical perspective, the question of what constitutes mathematical activity has been a matter of debate since the time of the Greeks. Part of the controversy revolved around the existence of mathematical objects.

For Plato, mathematics was a unique window onto epistemic truth in a privileged ontological relationship with reality. Later, Aristotle claimed that what was special about mathematics was its logical foundation and the nature of reasoning within it. These two positions encapsulate what can be taken as unique about mathematics; its objects and the interrelationships between them expressed in a rigorous and precise language.[1] The evolution of this language, and the philosophical disputes over what comprises the essence of mathematics are not our concern (see Ernest, 1994; Restivo, 1992). Our interest centres on meaning, and a crucial dilemma which emerges from even a cursory engagement with the philosophical is this: how far is mathematical meaning derived from its objects and their ontology and how far is it constructed through the linguistic domain by axioms from which conclusions can be drawn? (Otte, 1994, describes these as conceptual and formal activities respectively). Put crudely, we might ask whether mathematical meanings are derived from a metaphorical axis of referential objects, or from a metonymic axis arising from the formal relationships between them?

So philosophy offers us a first dichotomy, between mathematics as a self-contained formal system, or mathematics as a way of conceptualising the world: an opportunity to consider the place in mathematics of referential meaning, the practical, the 'real'. From an alternative perspective, we might ask a related question: is mathematics an object of study in its own right, a cultural form, or is it a tool for understanding the mechanisms of social and scientific life? Now it is clear that a simple answer to this question is that both are true: they are two sides of a coin. But this begs the question of the relationship between the two views, the relative weight which could and should be assigned to them. From a pedagogical point of view, it is the balance between these twin aspects which structures how meanings are ascribed by the learner to any mathematical activity – what must be stressed and what can safely be ignored.

In the following sections we take up different vantage points from which to consider the multiplicity of mathematical meanings, and to throw light on these two fundamental polarities. We begin by surveying major trends within the mathematics education community. After a vignette from our own research, we return to the literature in order to focus on the influence of context and setting on mathematical behaviour, a theme which emerges as salient in the preceding sections. We then consider the idea of 'street mathematics' in some detail, which will help us to clarify a framework for considering mathematical meaning which reconceptualises some of the fundamental questions and polarities we have discussed, and sets the stage for the theoretical ideas which we elaborate in the remainder of the book.

2. MEANINGS IN MATHEMATICS EDUCATION: A BRIEF SURVEY

The community of mathematics education is little more than 25 years old, depending on how its membership is perceived. In this short time, there have been wide swings of methodologies, realignments of theoretical frameworks, and occasional paradigm shifts. By tracing some of this history, we will see a fundamental shift from a focus on mathematical objects and how they are understood in the school population to concern with their construction and meaning. Our own ideas have evolved from this tradition, as well as from outside it; our purpose in surveying it is to understand this shift in perspective, its strengths and limitations, and to draw on this corpus of work to build a theoretical framework for understanding mathematical learning, conceived in its broadest sense.

2.1. *Meanings from mathematical objects*

Much of the research in mathematics education during the 1970's studied students' understanding of a range of mathematical topics selected from the school curriculum – that is, of mathematical objects (for example, reflections or negative numbers) and fragments of mathematical language (for example, 'if-then' arguments). Methodologies tended to be limited to the investigation of the individual student's grasp of mathematical content and data were largely restricted to measures of competence or, more usually, its absence. The main data collection technique was paper and pencil test, most often analysed by the simplest of statistical methods with little attention to 'background' measures – data were seldom if ever stratified according to school or student characteristics, nor account taken of topic exposure or teaching.

Thus the early days of mathematics education research were dominated by surveys identifying student errors. Errors which appeared to be widespread and consistent, termed 'misconceptions', became objects of experiments and 'treatments' were devised to effect their 'remediation'. Some indication of the strength of this perspective can be found in the proceedings of the International Conferences for the Psychology of Mathematics Education (PME). For example, of the 45 papers included in the proceedings of PME III (in 1979) all but three reported research in this paradigm. Investigations within mathematics education at this time remained remarkably untouched by concurrent developments in psychology which were increasingly taking account of the contextual determinants of task experience and deployment of skills.

By far the most influential research within this paradigm in the U.K. was the Concepts in Secondary Mathematics and Science study by Hart and her colleagues (Hart, 1981). CSMS investigated a range of topic areas,[2] within each of which, a total of several thousand 13, 14 and 15-year-olds were surveyed in over 50 schools using paper and pencil tests. Within each topic and age group, facility levels derived from the percentage of correct responses were recorded together with the most common errors. The descriptions of the misconceptions identified in this study had a major impact both nationally and internationally – they opened the eyes of teachers and researchers alike to a new appreciation of just how far children's views could diverge from those of mathematical orthodoxy.

One assumption underlying this and similar studies was that meaning was transparent – the questions in the tests reflected exactly the mathematics 'residing in' the topics under investigation and the written responses were reasonable manifestations of children's understanding of them. Context did not enter the picture: erroneous responses provided an accurate picture of the understandings ascribed to mathematical objects by children of different ages. One outcome of this research paradigm was the specification of 'hierarchies of understanding' which distinguished levels of difficulty, either within a topic layering mathematics according to its conceptual complexity, or across topics, labelling children according to their mathematical competence – a 'level 2 child', for example, was seen as capable of managing a set range of tasks over a variety of domains. These hierarchies became of particular interest to curriculum developers and policy makers.[3]

This uncomplicated and widespread approach was destined to be short-lived, for it contained within itself a startling irony. The studies manifested for the first time how student visions of mathematical objects were frequently at odds with those of experts. But in so doing, they cast doubt on the very premise upon which the research had been designed: the existence of an unambiguous relationship between mathematics and its understanding. Something more was needed.

2.2. Meanings from problem solving

The late 1970's and early 1980's gave birth to a growing disillusionment with the misconceptions paradigm. The treatment of mathematical knowledge was criticised: in misconceptions research, the objects of mathematics tended to be regarded as isolated fragments and performance in each topic studied independently of each other. The methodology itself was questioned: first, for its basic assumption of transparency which we have already mentioned; second, for its almost exclusive use of paper and pencil tests, responses from which could only paint a very partial picture of children's mathematical competencies; and third because of the rather limited interpretative power of the quantitative measures adopted and their analysis. In a retrospective survey, Schoenfeld (1992) marks as a turning point an article by Kilpatrick (1978) which compared the apparently rigorous but unilluminating 'scientific' methods in the U.S.A. with the more insightful, qualitative research undertaken by Krutetskii (1976) in the Soviet Union.

Thus research into students' mathematics began to shift from the simple assessment of problem solutions to concern with the strategies adopted during problem-solving. With this new focus came changes in methodology – interviews, observation and case study became more widely adopted. Projects proliferated which attempted to identify and to enhance problem-solving processes and their application. Later, attention turned to the assessment of problem-solving on the grounds that if it was not assessed, it would have little if any status in the eyes of teachers and students.[4]

Alongside the drive to develop fluency within the discourse of mathematical activity, there were parallel moves to assist students in becoming more competent users of mathematics. Students were asked to solve 'real problems' by modelling them mathematically, applying mathematical techniques and 'translating' the results back to the original situation (Silver, 1992). The problem solving movement[5] anticipated that, aside from the motivation derived from a contextualised problem situation, the investigation of the process of model building would shed light on the mathematical meanings brought to bear on any problem. Context was recognised as important, but it was seen primarily as 'a factor affecting performance', a variable to be manipulated or ignored if analysis became over-complex. In any case, the relationship of mathematical performance to context still appeared to be viewed as relatively unproblematic – mathematical tools could be simply applied and manipulated. As we shall see, this view proved inadequate as a way of making sense of learners' mathematical performance, let alone the nature of their mathematical behaviours.

Over time a new set of concerns began to come into view. First, concern was expressed at the disconnection of processes of 'scientific enquiry' from 'mathematical content': put bluntly, how can one learn to plan, monitor, or prove in the abstract? Recognition of this absurdity led to the acknowledgement of the essential complementarity between process and content and the importance of analysing the totality of a mathematical experience.[6] A second concern was that a problem-oriented approach which prioritises the 'tool' aspect of mathematics inevitably fragments the mathematical domain – connections within the problem (its setting, its less-obviously mathematical aspects) can easily take precedence over connections within mathematics. Thirdly, taking the problem situation as the arbiter of meaning is fraught with pitfalls, not least because the mapping between the mathematical and situational elements of a problem turns out to be highly ambiguous with respect to the mathematics deemed to be relevant, the aspects of the setting to be considered and the extreme sensitivity of problem-meanings to social and cultural parameters.

In the light of the first two concerns, approaches which emphasised process – especially at the expense of content – gradually gave way to the constructivist movement of the 1980's and 1990's with its stronger theoretical underpinning in philosophy and psychology. The third concern became largely superseded by the general shift in the social sciences to the more situated view of knowledge and cognition which we will discuss later in this chapter.

2.3. *Meanings constructed by the individual learner*

One spin-off of much of the research on misconceptions and processes was that it became clear that many children (at least in the U.K. and U.S.A.) constructed their own approaches to mathematical problems, and failed to make use of the approaches introduced in school (Booth, 1984; Hart, 1984). These 'child methods', although deemed incorrect from a mathematical perspective, turned out not to be merely random responses but rather sensible, even consistent ways to solve given problems. Frequently, child methods had their origins in successful methods which had been over-generalised; that is, the constraints necessary for correctness were not fully appreciated or were ignored. For example, consider the following child response:

$$\frac{4+3}{x+4} \longrightarrow \frac{\cancel{4}+3}{x+\cancel{4}} \longrightarrow \frac{3}{x}$$

One interpretation would be that the child was simply wrong.[7] A more constructivist interpretation would be that the child had extended the range of problems where fractions could be cancelled down by relaxing the constraint of multiplication.

The identification of child methods suggested that mathematical meaning interleaved mathematical objects and relations with contextualised nuances, stemming largely from prior experiences: quite how this interleaving works continues to be the subject of debate.[8] Worse still, with the more general growth of the constructivist paradigm, it became increasingly clear that child methods tended not to be discarded in the face of conflict or eradicated by teaching, but rather coexisted side-by-side with taught approaches (Fischbein, Deri, Nello, & Marino, 1985). This realisation brought forth a burst of energy which was both stimulated by and supportive to the constructivist movement[9] which has dominated the recent history of the mathematics education community. Its aim was simple enough to state, if not to implement: to find ways to build upon child mathematics rather than to ignore it.

It is unnecessary to describe constructivism in any detail here; there are numerous texts that admirably summarise its foundations and implications (PME, 1987; von Glasersfeld, 1995; 1991). Key elements of its educational application might be summed up as: student control over the direction of their own learning, recognition of the epistemological validity of student knowledge, the importance of fostering purposeful activity for the transformation of cognitive structures and the centrality of cultivating a sense of satisfaction and competence in learners.[10]

While we will not elaborate these elements here, one central facet of constructivism is germane to our discussion, and it is admirably summed up by Smith, diSessa and Roschelle (1993) who state: 'Constructivism demands that more advanced states of knowledge are psychologically and epistemologically continuous with prior states' (p. 147). The methodological corollary of this observation is that in trying to understand student responses, there is no alternative but to seek out prior meanings and identify the contexts where students' ideas and current understandings are productive. Noddings (1990) goes further and argues: 'constructivism may be characterised as both a cognitive position and a methodological perspective' (*ibid.* p. 7). Methodological constructivism involves clinical interviews over substantial periods of time to ascertain student approaches as clearly as possible; and it brings the teacher into focus – it is the teacher who must encourage the learner to reflect upon and conceptually debug intuitions, relate new knowledge to existing structures and widen the contexts in which mathematical processes can be applied. The influence of both these methodological tenets have been evident within mathematics education research of the last decade.

It is clear that constructivism is deeply influenced by a Piagetian model of learning – in fact Noddings goes so far as to suggest that 'there seem to be few epistemological advances beyond Piaget' (*ibid.* p. 9). For Piaget, mathematics is developed by interaction, it is neither given in reality nor inscribed in the mind, but comprises a synthesis of discovery and creation. This twin view fits neatly into the mathematics teaching/learning situation – teachers have to operate *as if* mathematical objects exist and need to be discovered, whereas children need to construct these objects for themselves. A primary Piagetian notion is that of a *scheme*, the result of the co-ordination of and reflection upon interactions with the environment, thus comprising the generalisable dimension in any set of actions after the particularities (including the objects on which it is applied) are set aside. A scheme is the primary instrument of assimilation – a mathematical object can be assimilated into a scheme, the representation so constructed being termed a *schema*. Thus, 'the use of a scheme results in the development of a schema to which the scheme is applied. By having to adapt to new objects, the scheme becomes more general and more flexible' (Nunes, Schliemann, & Carraher, 1993, p. 138).

Yet there is an unresolved dilemma: while the theory recognises the role of action and reflection in mathematical learning, there remains the question as to which dimensions amongst the actions are to be generalised? As Noddings points out: 'Students will construct but we want their constructions to be guided by mathematical purposes, not by the need to figure out what teachers want or where they are headed' (*ibid.* p. 16). This dilemma is at the heart of the constructivist programme. If students control the trajectory of their learning, how can meaning be constructed in a way required for the development of acceptable mathematical knowledge? More fundamentally still, there is a related epistemological question: if students are constructing for themselves, how shall we know when to count their constructions as mathematical?

2.4. *From enactive to reflective mathematical knowledge: spirals of abstraction?*

The tension between the autonomy of the student to construct knowledge on the one hand, and the authority of the knowledge domain known as mathematics on the other, is one to which we will return many times before we are finished. For it is at the intersection of these two, sometimes conflicting, trajectories that we will locate the creation of meaning which we seek.

It is relatively uncontentious to remark that children better construct mathematical meanings by using mathematical ideas in environments where they have clear functionality and purpose, by layering understandings in the course of use through the discrimination and generalisation of the essential structures. Others have made much the same observation. For example, the French school of mathematical didactics[11] stresses the essential interrelationship of action and conceptualisation together with the centrality of the problem. Vergnaud (1982) has gone so far as to define the meaning of a mathematical concept as in part derived from the set of problems to which it provides a means of solution.

However, there remains the need to interpret the process of mathematisation, the move from problem to mathematics, and it is here that a critical difficulty emerges. A classical view of mathematisation sees it as one of decontextualisation, a process of extricating the mathematics from the problem, removing it from action to cognition (see, for example, Vergnaud, 1987). This transition from direct knowledge, knowledge in action, to reflective knowledge, is generally described as the process of *abstraction* or among those with a more explicit Piagetian focus, *reflective abstraction* (Piaget, 1970). This does not simply involve the separation and retention of different qualities of objects but rather consists in recognising actions on these objects, reflecting on them i.e., projecting them on to the plane of thought, and finally integrating this new form into a fresh structure which is a reconstruction of the former.

The metaphors of ascent are all-pervasive. The French didacticians, for example, see mathematical concepts as 'progressing' from use to knowledge, through a cycle which moves from action to 'formulation' in a conceptual domain.[12] Laborde (1989) suggests that: 'the usual formulations of mathematical discourse require a certain level of acquisition of mathematical objects and relations: they must be sufficiently decontextualized and detached from students' actions... Conversely the requirements of formulation lead the individual to envisage mathematical objects in another way' (*ibid.* p. 35).

The process of abstraction has received considerable attention in the work of a subgroup of mathematics educators concerned with 'Advanced Mathe-

matical Thinking'.[13] The AMT group, crystallised around a Working Group
of PME, has published a volume (Tall, 1991a) which collects together an
overview of its approach and provides a useful starting point for examining
the idea of abstraction. For the AMT community, abstraction is a key component in the creation of a mathematical world populated by an interconnected
set of mathematical objects.

Just what *is* abstraction? Robert & Schwarzenberger (1991) describe abstraction as involving the recognition of objects and properties which not only
apply to the objects from which a generalisation is made, but also to any other
objects which obey the same properties. Dreyfus (1991) suggests that abstraction comprises a shift in attention involving mental reconstruction after which
the relationships between objects become central. Dubinsky elaborates the
notion of reflective abstraction by isolating five phases in this process: interiorisation, coordination, encapsulation, generalisation and reversal (Dubinsky,
1991, p. 103).

These formulations point to a broader theory which is current in the mathematics education literature. It starts from the observation that there is a
duality inherent in mathematical objects, a duality which emanates from their
appearance sometimes as process, sometimes as object. The list of workers
who have commented on this duality is substantial, and includes Freudenthal,
Douady, Kieren, Kaput, Dubinsky, Greeno, Tall and several other notable
researchers in the field. Sfard (1991) has added to our understanding of this
duality by distinguishing two views of mathematics: as a static set of abstract
objects – a *structural* conception; and as a sequence of processes, algorithms
and actions – an *operational* conception. She suggests that despite the essential complementarity of the two conceptions, there is a deep ontological gap
between them with the latter 'progressing' to the former through three stages:
interiorisation (essentially the Piagetian use of the term), *condensation* where
sequences of operations are thought of as a whole, and *reification*, where 'the
new entity is *detached* from the process which produced it and begins to draw
its meaning from the fact of its being a member of a certa in category' (p. 20:
emphasis added).

Clearly, abstraction is not the same as generalisation. Tall (1991b) argues
that the former is merely an extension of the familiar while the latter demands
a new vantage point defined by a set of axioms and necessitates 'a massive
mental *reorganisation*' (*ibid.* p. 11, emphasis in original). Dreyfus similarly argues that neither generalisation nor synthesis (two processes that, he
suggests, form a prerequisite basis for abstracting) make the same cognitive
demands on students as abstraction. These contributions provide important
insights into the nature of mathematical activity, and how it can be distinguished from other forms of academic discourse. We cannot do justice to the
details of the entire position here, but we believe they share an agreement that
mathematics is hierarchically organised into spirals of abstraction which are

recursively reproduced – the stage of reification, in Sfard's terms, being the starting point of the next cycle.

We have two concerns with this theory. First, the social dimension of learning is ignored; we postpone consideration of this point until later in the chapter. Second, there is a slippage from epistemology to psychology. This slippage is nicely illustrated by Sfard & Linchevski (1994) who point to students' 'natural propensity for an operational approach' (p. 224) and argue that this should be exploited by an appropriate pedagogy. Now we would not argue that such an approach may often be appropriate. Neither can we disagree that many, perhaps most, children derive their understandings of mathematical objects in this way, by an operational approach to mathematical objects which only subsequently become mentally 'reified' as objects.

There are, however, three points with which we take issue. First, we are sceptical that this approach is independent of the mathematical objects concerned. For example, Sfard and Linchevski illustrate their theory by arguing that 'abstract' objects like $\sqrt{-1}$ and the function $3(x + 5) + 1$ are the outcome of 'reification'; yet while we can appreciate what is intended with respect to the latter function, we have difficulty in understanding how the abstraction of $\sqrt{-1}$ as a reified entity emerges from the process of extracting the square root of -1. Second, we wonder whether the argument holds equally for all mathematical domains; what is the geometrical analogue of (algebraic) reification? Third, and most critically, we are unconvinced that this propensity is 'natural' any more than it is (was?) 'natural' to study fractions before decimals. In its present form, the theory of reification leads to the inescapable conclusion that the process of understanding is a one-way, hierarchically-structured and uphill struggle (literally: hence 'a jump from operational to structural mode of thinking means a transition [...] from the foot of a mountain to its top' (*ibid.* p. 203).

Learning and mathematical knowledge climb together to ever more-rarefied heights of abstraction, leaving, inevitably, more and more casualties in their wake. As Sfard herself has asserted: 'Being capable of somehow 'seeing' these invisible objects appears to be an essential component of mathematical ability; lack of this capacity may be one of the reasons because of which mathematics appears impermeable to so many 'well-formed minds' (Sfard, 1991, p. 3). The implication is clear, and echoes the sentiments expressed a decade earlier: 'It is *impossible* to present abstract mathematics to *all types of children* and expect them to get something out of it' (Hart, 1981, p. 210: emphasis added).

Now the problem is not merely one of characterising children into 'types', as having or lacking 'capacities'. It is a problem of meaning. Where can meaning reside in a decontextualised world? If meanings reside only within the world of real objects, then mathematical abstraction involves mapping meaning from one world to another, meaningless, world – certainly no simple task even for those with the 'capacity' to do it! If meaning has to be generated from within mathematical discourse without recourse to real referents, is this not inevitably impossible for most learners?

We have reached an educational impasse. On the one hand we are faced with the formal and decontextualised – mathematical, abstract; on the other hand, informal and contextualised – everyday, referential. Mathematicians, we include ourselves, can surely recognise the power of the former; yet immersion in the latter seems to deny many access to anything that can count as mathematical.

Mason (1989) points to a way out of this difficulty, by characterising mathematical abstraction as the result of a 'delicate shift of attention'. This formulation locates abstracting as a flowing process where a sense-of (Mason's word) constructing a mathematical expression is retained even after the expression becomes an object. On the level of the individual learner, this description is helpful and we try to build on it in trying to resolve the more general impasse posed above.

Our answer will consist of several parts. We will need to look more carefully at traditional dichotomies, through socio-cultural, as well as epistemological and psychological windows. and disaggregate their components: does, for example, the idea of the formal necessarily imply decontextualisation? Are formal and informal polar opposites? Second, we will need to disaggregate the idea of referent from the idea of meaning; it will turn out that meaning can derive from other than 'real' or referential sites. Conversely, loss of referent is not coterminous with loss of meaning. Third, we will need to look more closely at the notion of abstraction itself; paradoxically it will help to set up yet another dichotomy, between abstract and its apparent opposite, concrete.

Resolution of these issues will have to wait: indeed, it will take up most of the remainder of the book. In this chapter, we continue to lay the foundations for our theory, by addressing what we and other workers in the field have to say about notions such as context, formalisation, and abstraction. Before we do so, we will describe a study we undertook a while ago, which gave us pause for thought at the time and provided us with many of the questions we now attempt to answer.[14]

3. VIGNETTE: THE N-TASK

We were working with a group of seven thirteen-year-old children, all of whom had been learning Logo for some considerable time. We began to introduce structured tasks into our activities in order to explore more carefully the kinds of meanings constructed by the children during their work, and to try to understand how their thinking was influenced by their computer interactions. The mathematical domain we chose was the idea of ratio and proportion, a topic which had been the object of much research and was widely regarded as difficult for children of this age.

We devised a task the first part of which consisted of a sequence of letter N's with vertical sides 150, 350 and 100 units and with an angle of 45°. The children were asked to predict the length of the diagonal of each N, write down a Logo procedure for each N on paper and try out their procedure on the computer with the normal Logo-turtle commands. The sequence was repeated with 30° N's, after which the children explored their own procedures by designing and constructing a pattern of their choosing. Finally, they were offered a new problem: to write a procedure which could draw proportional N's with a 30° angle. The intention was to encourage the children to generalise and express in Logo the invariant relationships which they perceived between the lengths and angles of a 30° N, to name this new object and to test it out in a social context by reference to each others' work. The final part of the task consisted of the following problem:

You have all designed your own patterns.

There is a procedure – SUPEREN – which could be used to draw all of the N's in *everybody's* pattern.

Can you write SUPEREN?

Your SUPEREN will be tested on somebody else's pattern.

We should point out that, provided that one accepts the rules of our game – i.e., that the figures should be drawn 'accurately' and that guessing the lengths and angles was not 'sufficient' – the task was rather demanding. As well as recognising the angle relationships, a complete (official, mathematical) solution would involve recognising that a multiplicative and not an additive relationship existed between BC and AB and that the relationship between AB and BC depended on the size of the angle ABC (see Fig 2.1).

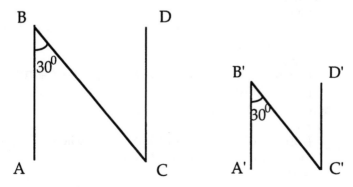

Figure 2.1. Two proportional 30° N's.

What did we expect the children to do? It certainly made sense to draw
on the many research studies which existed (all had used pencil and paper),
and set out our predictions for children's strategies. At this stage, we, like the
authors of most studies at that time, could only assume that these strategies
would be independent of the medium in which they would be expressed (a
fact which, interestingly enough, is masked by the adoption of strictly mathematical language to describe them). In brief, we classified likely strategies
as follows:[15]

A. Addition with constant differences – where the relationship within
   the ratios is computed by subtracting one term from another and the
   difference applied to the second ratio:
   i.e., If $BC = AB + x$, then $B'C' = A'B' + x$
B. Addition with adjusted differences – where a relationship within a
   ratio is established and then extended to a second ratio by addition.
   Piaget et al. (1968) calls this 'preproportionality'. The amount added
   is 'adjusted' using arguments which vary in level of sophistication as
   follows:
   i) Visual adjustment of the difference between $B'C'$ and $A'B'$ –
      i.e., the bigger the 'N' the bigger the increment.
   ii) Increasing adjustments to the difference between $B'C'$ and
      $A'B'$; where if $B'C' = A'B' + x'$, then x' is a discontinuous
      function which takes different constant values for different
      lengths of AB.
   iii) $A'B'$ is multiplied by a reasonable scale factor and the answer
      adjusted by addition or subtraction to obtain a value for $B'C'$.
C. Proportion by repeated additions – a correct though cumbersome
   strategy. For every AB (or part of AB) in $A'B''$, an amount equal to
   BC (or part of BC) is taken. These are then added together to give
   $B'C'$.
D. Proportion by addition:
   i) Functional
      If $BC = AB +$ (fraction of AB), then $B'C' = A'B' +$ (same
      fraction of $A'B'$)
   ii) Scalar
      If $A'B' = AB +$ (fraction of AB), then $B'C' = BC +$ (same
      fraction of BC)
E. Proportion by multiplication:
   i) Functional
      If $BC = AB \times k$, then $B'C' = A'B' \times k$; where k is spotted by
      the student and not obtained by division.
   ii) Scalar
      If $A'B' = AB \times k$, then $B'C'' = BC \times k$; where k is spotted by
      the student and not obtained by division.

Let us summarise what happened. Only one student appeared initially to use addition with constant differences. Six of the seven children eventually used one of the proportional strategies classified as C, D or E above; that is they recognised that BC is longer than AB by an amount proportional to the length of AB. We were mildly surprised: from our review of the ratio literature, we had expected many if not most of the children to use an addition strategy; still, a sample of seven is hardly conclusive! But we had a pointer to something more interesting. We were convinced by our observations that the children had seen that the relationship between the lengths *had* to be multiplicative – they had recognised the proportionality of the situation in some implicit way – so their struggle was to find a way to express this. To better understand this we had to look more closely at how some of the children's strategies unfolded in the course of their activities.

We start with Nicola. Nicola adopted an adjusted difference response (category B) but was clearly very close to developing a proportional one (category C). This she used later, explaining that *'For a 45° N, every 50 (in AB) means an additional 25 along BC'*. Note how she recognised the dependency of BC on the angle – she knew that if the angle changed the additive constant must also change. Nicola's SUPEREN procedure looked like this:

```
TO SUPEREN :FO :RI :OT
FD :FO
RT :RI
FD :OT
LT :RI
FD :FO
END
```

The procedure encapsulates some of the invariants of the task in symbolic form, notably that the vertical sides and the opposite angles of N's are the same. But now Nicola's awareness of the relationship between OT and FO became evident: when she called her procedure she only chose values for her inputs which produced a correct figure. In crude terms, she knew what the relationship was even though she appeared unable to express it in the code.

Paul and Noel started with an intuitive approach (category Bi); for example, Paul wrote as an answer to a question about the length of a diagonal of a 45° N: *'B'C' is 510 units because as it is 45 it has further to go'*. They then both predicted increasing adjustments to A'B' (Category Bii). These were inconsistent across different size (but same angle) N's. Finally the boys were able to make a Category Di response to predict the diagonal length of any 30 degree N:

$$B'C' = A'B' + \tfrac{1}{6} \cdot A'B'$$

They expressed this in Logo in the following SUPEREN procedure:

```
TO SUPEREN :F
FD :F
RT 150
FD (:F+:F/6)
LT 150
FD :F
END
```

It seems that they had made some transition from Category B to Category D responses: but how? Noel's work illustrates what happened. In constructing an N where AB was 150, Noel obtained a value of BC of 175 by iteration – by guessing and testing visually. When AB was 350, his first guess at BC was 400 (a category Bi response). Noel then thought very hard, turned to one of us, and asked if he could use the computer 'to do a sum'. He tried different operations on 350, trying to get a number which, when added to 350, would give him 400. After several attempts he muttered '*It is something to do with 6*', and tried 350/6 – giving 58.3333. Despite being rather disquieted by the decimals, Noel decided on the following answer '*I had added one sixth of 150 before so I add one sixth of 350*'. This he generalised: '*You take a one sixth of AB and add it to itself to make BC*'. When he constructed his Logo program he went further: he added a fraction of an input to itself (in the line including :F+:F/6). So close inspection of the evolution of this strategy reveals how the expression of the mathematical relationships was mediated by the structure of the environment – from an operation on numbers, on sides of his N, and finally on a variable.

Noel and Paul were even able to describe how their SUPEREN procedure could be seen as a particular case of a more general procedure to draw *any* N. Paul remarked: '*You will have to have a :A then a times or divided by something that will go instead of a divided by 6*'. This idea was taken still further by David who spontaneously came up with a way of designating a 'something that will go instead of a divided by 5' by calling it :FORMULA in the following SUPEREN procedure:

```
TO SUPEREN :SIZE :D :FORMULA
FD :SIZE
RT 180 - :D
FD :SIZE +:SIZE /:FORMULA
LT 180 - :D
FD :SIZE
END
```

David's program is remarkable: it provides evidence that he was searching explicitly for a generalisation of his division by 5 for the 30° N, to division by a variable : FORMULA. Although there is no doubt that David was thinking about : FORMULA as an integer (say 5 or 6 – both inputs which he tried as values), the formalisation in Logo in fact *achieved a much greater generality* which could serve as a hook towards more precise relationships. Similarly, the formalisation of a functional proportional strategy by Paul and Noel, first conceived in terms of integers only, became immediately generalisable to a larger range of numbers – and the children recognised this. So something interesting is occurring here: a generality and a new object which is implicit yet precise, vaguely appreciated yet rigorously expressed.

But we are not yet finished with Paul and Noel. Fortunately, an incident which had occurred some weeks earlier, allowed us further insight into the complexity of the situation. At this (earlier) point, Paul and Noel had decided to write an interactive quiz in Logo and wanted to write a program for its title, which was, appropriately enough, called 'Q U I Z'. They adopted a modular approach building separate programs for each letter and separate interfaces, planning their moves on paper. In direct mode (i.e. one command at a time) they drew a Q (a square of side 100 and a small tail of length 20; see Figure 2.2). After a little discussion with us, they made their procedures variable and with no apparent difficulty substituted :S for 100 in their program for Q. When they arrived at the 20 in FD 20, their first inclination was to use another input and to subtract but when they came to write their program at the computer choosing subtraction as a possible operation simply did not occur to them: '*FD what? Oh well, it was FD 20 and :S is 100 so it's FD :S divided by 5*'.

Figure 2.2. Paul and Noel's Q.

Later, when they came to substitute for BK 40, they continued this line of argument: 'That BK 40 is twice as big as the FD 20 – so you have to divide the 100 by half as much.' This was expressed in Logo as a division of the input by 2.5 so their program for Q was as follows:

```
TO Q :S
REPEAT 4 [FD :S RT 90]
RT 90
FD :S
RT 25
```

```
FD :S / 5
BK :S / 2.5
FD :S / 5
LT 25
END
```

Finally, when they came to build their procedure for Z, they went straight
into the editor and typed:

```
TO Z :S
RT 90
FD :S
BK :S
LT 45
FD :S * 2      ;;the diagonal line is twice the length of the horizontals
LT 135
FD :S
END
```

They tried out their Z (using an input of 100) and decided the 'slanty' line
was too long. After experimenting with various numbers 1.2, 1.3 (they were
happy to use decimals), they eventually decided that 1.4 worked the best!
Working exclusively on the computer, Paul and Noel had *taken for granted*
the multiplicative relationship between the two lengths in question and arrived
at a fair approximation for the square root of 2.

We have two observations on this data. First, the interplay in the N task
between pencil and paper responses and computational ones, forces our attention on the ways in which meanings are mediated by the setting, and in
particular, on the medium available for expression.

Figure 2.3 is a schematic map of the strategies adopted by the children in the
sample in terms of the previously identified paper and pen strategies: we have
denoted by an arrow typical trajectories of children's approaches. We have
indicated where we can see equivalent strategies instantiated in the different
media (pencil-and-paper/computer). In the first place, it becomes apparent
that there is no computer behaviour analogous to the additive strategy. A
second observation concerns the lack of connection between the addition with
constant and adjusted differences and proportional strategies in the paper and
pencil scenario. There was no conceptual route connecting these strategies,
other than that mediated by the computer (through iterated differences or
some other quasi-algebraic expression such as :S + a half of :S). Here is our
first clue to the two process we seek to understand: first, what the computer
can bring to the 'understanding' of proportionality (or at least our observation
of a behaviour which seems to exhibit such understanding); and second, how
such understandings are mediated by the child-computer interaction.

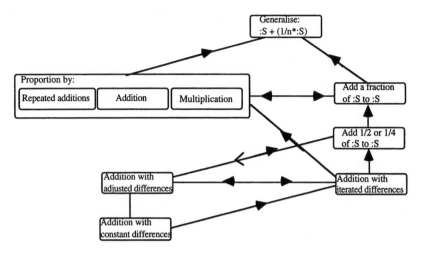

Pencil and paper strategies   Computational strategies

Figure 2.3. A map of student strategies in the SUPEREN task. The arrows denote common trajectories of children's approaches. The open arrowhead indicates an inconsistently observed behaviour on the children's part.

The strategies which Paul and Noel adopted for their Z task are much more straightforward to exhibit since they worked exclusively on the computer. As illustrated in Figure 2.4, they bore marked differences in comparison with their previous N strategies, exhibiting little connection with pencil and paper approaches (see Figure 2.4):

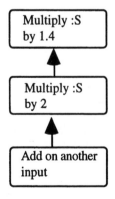

Figure 2.4. Computational strategies for Paul and Noel's Z.

We do not intend to delve deeper into these data, particularly as they represent only an early indication to us of some interesting phenomena, and suggested lines of enquiry some of which we will take up in later chapters. We focus on two issues. Our first point is methodological. By synthesising careful task analysis with detailed observation, we could illustrate that what children did was intimately bound up with the specificities of the task, and with the medium of expression available. It makes little if any sense to ask what Noel and Paul knew, rather than what they knew when, where, and how.

Our second point is less concerned with outcomes and more with cognitive process. The data illustrate that unexpected levels of generality can emerge from computer interactions, and that formalisation in quasi-algebraic terms – in the form of computer code – can become a means of thinking about and expressing relationships, rather than merely summing up already-understood relationships. Quite how this works will involve piecing together a rather complex puzzle of theoretical and empirical material. Whatever the explanation, there were startling differences in approach mediated by the differences in the settings: and it is to this influence on setting that we now turn our attention.

4. THE INFLUENCE OF SETTING ON MATHEMATICAL BEHAVIOUR

The acknowledgement that setting is intimately bound up with performance on mathematical tasks precedes the widespread acceptance of the situated view of cognition. In fact, we can easily underline this point by reference to a sequel to our N-task, in which we gave our case-study children a pencil-and-paper test, based on the CSMS test for ratio (Hart, 1985). One item is reproduced in Figure 2.5 below:

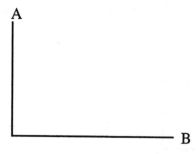

Work out how long the missing line should be if this diagram is to be the same shape but bigger than the one above:

.........cm   B

Figure 2.5. An item from the CSMS study.

Not surprisingly, this is a difficult question. Its facility level (12.5% for children aged between 11 and 12) placed it at the bottom of the CSMS hierarchy. Quite apart from the geometry of the situation, the solution involves the recognition of a multiplicative relationship and one that requires a non-integer multiplier (the ratio is 5:3). On this item, none of our experimental group of children produced the 'correct' answer of $\frac{10}{3}$, although three produced an answer of 3.5 which was judged within the limits of acceptability by the CSMS team. David, easily the most advanced thinker of the group both mathematically and in programming terms gave the answer 3. Paul left the answer blank.

How can this poor performance be matched with the sophisticated approaches displayed in the Logo environment? It illustrates what a considerable body of literature in mathematics education continues to establish: that is, it is impossible to isolate a mathematical 'concept' from its setting. Crucial influences of setting range widely, from the wording of the problem to the tools and resources available.

Mathematical competence does not come neatly packaged. For some considerable time, the limitations of a strictly Piagetian account of development – which presupposes invariance of understandings across settings – have become apparent. Criticism has come from an epistemological point of view which stresses the social construction of knowledge, as well as from a perspective of cognitive development which recognises that social factors are crucial to knowledge acquisition (see, for example, Doise, Mugny, & Perret-Clermont, 1975; Hughes, 1986). We look a little more closely at some of this literature below.

4.1. Problematising children's performance

Early limitations of the non-contextual approach emerged with the realisation that the paper and pencil mode of testing performance might not reveal a 'true' picture of a child's mathematical competence and that answers given might not provide the transparent window on to understanding that was previously assumed. How do children interpret the questions? How do they perceive the role of the experimenter, or the goals of the tester? How do the questions suggest certain strategies and answers?

A few examples taken from the Assessment of Performance Unit (APU) in the U.K. will serve to illustrate some of the issues. One of the approaches adopted in their paper and pencil surveys was to ask questions involving 'the same mathematical operation' but in different settings. The data showed conclusively that changes in setting produced quite dramatically different facility levels. For example, the APU (1986, p. 836) showed that at age 11, the item *4.5 + 0.5 = ?* was answered correctly by 63% of pupils. In contrast, the question: *John saved £3.70 and then his mother gave him £1.50. How much did he have in all?* was answered correctly by 82%. It is worth noting that the conventional wisdom concerning the relative difficulty of 'word problems' seems reversed here; a possible explanation being that money problems are so well-grounded in children's experience that such difficulties are 'overcome'. But that is not the point: the question was to try to understand what students knew about the addition of decimal numbers.

Other data from the same studies similarly pointed to the fragility of students' knowledge when viewed exclusively through the window of pencil-and-paper tests. The phrasing of a question could be critical: 'How many halves are there in $2\frac{1}{2}$?' had a success rate 30% more than '$2\frac{1}{2} \div \frac{1}{2}$' at age 11, and 29% more at age 15. Using 'of' instead of a multiplication sign when fractions were multiplied substantially improved performance at both ages (*ibid.* pp. 840–1). Similarly, the presence of diagrams or the use of non-standard conventions influenced performance.

In an early study, Nesher & Tecibal (1975) showed that the use of verbal cues can improve performance (such as 'more' means 'add') or lower performance (when 'more' does *not* mean 'add'). Unfamiliar words, whether important for mathematical solution (such as 'product' or 'isosceles'), or not (such as questions related to cricket), and the way in which test items are administered, all affected performance in ways that might depend on individual, group or cultural differences[16] (see Murphy, 1980).

The evidence – we have only illustrated a tiny fragment here – was clear. Contextual cues could help or hinder depending on their interpretation by the child, and this interpretation is not simply an artefact of methodology. Yet it is important to understand just how limited this observation is. It recognises context as a crucial facet of understanding, but in so doing, it regards context as 'problematic', a variable to be controlled for, or one whose 'effect' is, at best, to be acknowledged.

Recognition of the inadequacy of this position led to a shift within mainstream mathematics education research in the late 1980's towards a perspective incorporating broader analyses of social and cultural factors in any consideration of the construction of mathematical meaning. Some studies concentrated on the social constraints of the classroom, showing how mathematics constructed in classrooms tended to degenerate into linear accumulation of tried and tested routines: 'The mathematical logic of an ideal teaching-learning process ... becomes replaced by the social logic ...' (Bauersfeld, 1988, p. 38). Other studies pointed in equally promising directions, and began to investigate 'another mathematics' transacted out of school. So, largely but not entirely spurred by the necessity of understanding the influence of context – rather than simply allowing for it – research began to take account of the methodologies from sociology and anthropology leading to '...the evolution of over-arching frameworks, such as cognitive apprenticeship, that deal with individual learning in a social context' (Schoenfeld, 1992, p. 348). By far the most elaborated expression of this attempt to understand the role of context in understanding so far, has been undertaken by those who have positioned themselves within the tradition of situated cognition.

4.2. *Into the cul-de-sac: situated mathematical expression*

Most people are quite unaware that there is a mathematical dimension to their professional or everyday activity, even when they are involved in relatively sophisticated numerical or spatial activities (see, for example, Wolf, 1984). Instead, it seems that the mathematicians-eye-view of the world, in which there are evident structures which underpin activity-in-settings, is the exception rather than the rule.

The recognition that knowledge is bounded by setting has been most elaborated by the now-celebrated work of Jean Lave and her colleagues. Perhaps the most evocative and simple finding on which her theoretical ideas are based is her observation that calculations which shoppers in a supermarket performed were almost 100% correct; yet when the same shoppers were asked to perform the 'identical' calculations with paper and pencil, performance fell to about two thirds of that in the supermarket setting (see Lave, 1988). The quotation marks are evidently needed, if only because for the shoppers, the calculations were clearly *not* identical.

Lave's explanation for such disparities is part of a broader analysis in which she shows that the setting itself creates problems and structures its own solutions. In the supermarket, it makes sense to use strategies far removed from 'mathematical' practice, but which may resemble it: adding in a 'best buy' situation may be perfectly sensible while shopping, but evidently wrong in the context of pencil and paper word problems. The key point is that people construct solutions in the course of action, and that these solutions are structured by their activity. People avoid doing what might be classified as school mathematics in the supermarket not because it is too hard, but because the practice of supermarket shopping carries with it its own discourse, its own meaning-making mechanisms. It would be too simple, however, to conclude that there is nothing that passes for mathematical in shoppers' activities. The point is that when shoppers *do* use mathematics in the supermarket, it is supermarket mathematics, a mathematics made possible through the resources of the supermarket.

Mathematics of this kind differs in substantial respects from the kinds of mathematics commonly taught in schools. In the first place, arithmetical activity is part of an organic process where the focus is not on the processes of solution but rather on '...dissolving problems... making them disappear into solution within ongoing activity...' (*ibid.* p. 120). Second, calculation is less likely to provide the overarching rationality with which numbers are commonly endowed: decisions are made on the basis of other-than-arithmetic criteria: as Lave nicely observes: 'Calculating activity exists, but formal solutions, boxed products of calculation, are more often built into setting and activity or used as vehicles for the expression of feelings about rationality, than for its implementation' (*ibid.* p. 176). Situated mathematics has no need for universal laws, consistency, generality – it is concerned with getting the job done using whatever affordances are available.

These insights spread rapidly into the community of mathematics educators and were adopted implicitly or explicitly by many disparate groups all of whom acknowledged the centrality of context and culture. One such group grew around *ethnomathematics,* a continuing programme of investigation into the relationship between mathematics and the cultural traditions of a given population – be it in the industrialised or developing world (a useful starting point for this literature is Keitel, Bishop, Damerow, & Gerdes, 1989).

Alongside the ethnomathematical trend, there were more explicit attempts to analyse mathematics education along a political dimension. In his seminal book, *The Politics of Mathematics Education,* Mellin-Olsen (1987) opened the eyes of the community to the ways mathematics could be used by students as a means to express meaningful aspects of their lives. Later, Skovsmose (1994) spelt out the philosophy underpinning this position. We do not intend to summarise this trend here, but we should at least appreciate how far the political (sometimes referred to as *critical)* mathematics educators have moved the locus of investigation in the field: context has been moved to centre stage, activity has become the focal point for an analysis of the ways in which mathematics might empower and extend human action, mathematics is mobilised as a cultural resource in activity.

This shift prefigures an important, perhaps crucial broadening of the conception of what it means to engage in mathematical activity. We might choose to take issue with this approach – this degree of breadth runs the risk of mathematical imperialism, of subsuming under the title of mathematics any activity onto which *we* can impose a mathematical point of view. That is a real danger. But the alternative represents a far worse risk: of erecting a too strict and impervious boundary around what is, and what is not mathematics. This we are reluctant to do: indeed one of the projects of this book is precisely to show how the boundaries which have traditionally been built around the concept of mathematics are (and, although we do not consider it, have always been) relatively fluid, and that it is possible to think about other than official mathematics as mathematical, in a way which simultaneously democratises access to the field, yet does not dilute the mathematical enterprise.

Mathematics-in-activity is characterised by its mobilisation-in-use: its meanings derive from the need to solve a problem, or to achieve a specific outcome. The mathematics, just like any facet of activity, is not an object of conscious reflection, it is part of the action. This insight points to the abuse of language inherent in the idea of mathematics-in-activity: the activity is not constituted by its different elements (say, mathematical, physical, social) but by the dynamics of the activity as a whole. Mathematics-in-use may be quite different from formal mathematics, but from a pedagogical point of view, a crucial question is to define a relationship between them. Are they separate, united only by the use of the same noun by some (but not all) authors? Or are they, as some philosophers might have us believe, part of a unified whole, two sides of the same mathematical coin? If this is the case, how are the two sides related?

We might be content to consider informal mathematics, mathematics-in-use, as a starting point for more formal mathematical learning: Keitel (1986) refers to it as pre-mathematics. But we will still be left to delineate the boundaries of each, and to ask whether the former contains the seeds of the latter. Lave's discussion of cognitive tools offers help here. She argues convincingly that when 'tool' is used as a metaphor for knowledge-in-use across settings, it is inevitably assumed that there is no interaction between tool and situation and that a tool can be applied across a range of settings, through the process referred to as 'transfer'. Thus, in transferring across situations, the essence of the tool is assumed to be a passive entity. Mathematics can be taken as a classic example of this approach – mathematics as a decontextualised tool to be 'mapped' apparently unproblematically onto contextualised problems. Lave rejects this clear dividing line between 'abstract, decontextualised' knowledge, and immediate 'concrete, intuitive experience'.

Her position is now widely popular as part of the (mainly) North American critique of functionalist theory. Resnick, (1991) for example, argues that: '...every cognitive act must be viewed as a specific response to a specific set of circumstances. Only by understanding the circumstances and the participants' construal of the situation can a valid interpretation of the cognitive activity be made' (p. 2). This strong situated view has had considerable influence on the theoretical and methodological thrust of mathematics education research. Yet while Resnick's second, methodological, assertion seems eminently reasonable, the first raises a formidable challenge to the very essence of mathematics. For mathematics is the science of structure and pattern, the study of relationships between relationships; and to that extent at least, it cannot be immutably tied to 'real' referents, 'situations' whose meaning is contingent upon the particularities of setting: at least not in the traditional sense of 'real'.[17] The problem becomes clearer if we state a mathematical corollary of the strongly situationist view. Abstract and real are opposites: if it is real, it cannot be abstract and therefore not mathematical. If it is abstract, it cannot be real.

One solution is simply to define out of existence some or all elements of situatedness in relation to universal mathematical processes. For example, Lancy (1983) has argued that 'Grouping, categorising, generalising etc. represent a fundamental human need every bit as basic as the need to eat, to drink, or to socialise' (p. 64), and that these fundamental needs constitute mathematical activity. Bishop (1988) elaborates this position extensively by identifying six activities (counting, locating, measuring, designing, playing and explaining) which he proposes are central to the development of mathematics and are universal across cultures. By attempting to isolate pan-cultural facets of mathematical ontology these approaches aim to respect both cultural diversity and the epistemology of mathematics. Yet it is open to question whether they do so at the risk of losing sight of the specificities of mathematical knowledge. Is there an alternative perspective which respects the epistemology of mathematics and the situated way in which the learning of any knowledge domain must take place?

To answer this question we need to turn our attention temporarily from the role of context, to the dichotomy between formal versus informal thinking. We will try to put in place a foundation stone which clarifies the relationship between these two poles, and the related issues of representation and expression which will form important parts of the conceptual framework we are aiming to build.

5. STREET MATHEMATICS

There is a growing body of work which has investigated mathematics in a range of informal settings; in buying and selling (Carraher, Carraher, & Schliemann, 1985), basket making (Gerdes, 1988), packaging (Harris, 1987), building (Mellin-Olsen, 1987), and weaving (Fasheh, 1991). One of the most elaborated attempts to clarify the relationship between formal and informal mathematics has been made by Terezinha Nunes, Analucia Schliemann and David Carraher in their studies of 'street' mathematics which have now been integrated in Nunes et al (1993). They base their theoretical framework on a series of ingenious studies undertaken in Brazil over a period of some ten or so years where they compared mathematics used outside school in a variety of settings (street sellers in markets, fishermen pricing shellfish, foremen on building sites, carpenters and farmers) with performance on 'equivalent' mathematical tasks embedded in a school context (usually with the requirement that answers were written). The researchers used a mixture of methodologies to probe the thinking of the participants, starting with *in situ* anthropological observations to lay bare the mathematical ideas within the practice, often supplemented by task-based interviews, and occasionally followed up by experimental studies.

5.1. *Meaning mathematically in the street and in school*

The most obvious finding of the Brazilian studies was the marked differences in performance and strategy in the face of the 'same' mathematics in paper and pencil settings and in situ. As we saw in Lave's studies, the use of the word 'same' needs some elaboration: the tasks were structurally the same, but clearly differed from a psychological (or social) viewpoint. The Brazilian findings are strongly convergent with, but considerably more general than the findings we reviewed earlier, concerning the diversity of performance in school-based tasks.

Nunes and her colleagues explain their findings by observing that 'understanding an activity requires considerations about form, attitude, goals, and relationships between the activity in question and other activities' (*ibid.* p. 128). They suggest that the major difference between street and school mathematics is that the latter is separated from meaning and proceeds simply on the basis of rules and syntax, whereas the former preserves and exploits situational meaning in generating problem solutions:

> As the situational aspects are lost in written representation, syntactic rules become the focus. Conversely, oral arithmetic maintains in focus the meaning of the situations, and less attention is given to the syntactic rules.
>
> (*ibid.* p. 145).

Now the use of the word 'meaning' is problematic here. Nunes' observation that the presence or absence of situational information provides a key to understanding differential performance is important. She is using the idea of meaning to index situational referents: it is true that situations can confer meaning, but real referents are not necessary for such meaning to be established. As we shall see, it might be more helpful to consider meanings as derived from mathematical as well as 'real' settings: if that is the case, we could reformulate Nunes' observation by saying that the differences in performance reflect the difficulties children (and adults) have in coordinating one set of meanings with another.

While the Brazilian findings resonate with those of Lave and her colleagues, the former have centred their attention much more on how meanings are constructed in settings, and in particular, on the notion of representation. Let us take as an example, Nunes' observation on the use of a monetary system, a classic case of an everyday setting in which logicomathematical concepts are embedded (see Nunes, et al., 1993). Money activities routinely involve representing totals as constant, even though the coins which go to make up the total are of different values: in other words, there is a principle of 'additive composition of money' understood in the oral, money context. By contrast, calculation with pencil and paper normally involves additive composition of, say, tens and units rather than different coins of unit, five, ten and fifty units:

> Both representations respect the principle of additive composition, but the representation of the money total may *evoke* this principle much more clearly than does a written number. Perhaps the power to evoke the sense of the situation is one of the reasons for the discrepancies in performance we observed between street markets and schools
>
> (*ibid.* p. 144, emphasis in original).

Here is the crux of the argument. Some situations are structured in such a way that the underlying mathematical relations-in-use are more likely to arouse sense-making connotations which – in Nunes' terms at least – means connotations derived from everyday experience and activity. Moreover, 'In the written forms of representation typically used in school, the meaning of the situation is purposely not represented. It is by *not representing the referents* that it becomes possible, for example, to multiply across measures – that is, multiply money by number of coconuts and find a result in money' (*ibid.* p. 145, emphasis added).

We will consider this point later. Before we do, we need to look at other, equally important aspects of the Brazilian studies. Foremost among these is the observation that street mathematics solutions were not – as might have been expected – simply confined to routines habituated in the work context. Nunes and her colleagues claim that both conceptual and procedural knowledge may follow from practice, and conceptual schema clearly surpass the procedures used in everyday practice. The fishermen, for example, were able both to invert the calculating strategies they normally used at work and to utilise them in other domains within which they were not usually applied; still more interesting was the finding that the level of schooling had no effect on whether the knowledge developed outside school could be used in new ways. In sum:

> ...representation of the particulars of the situation does not imply that the subject is restricted to understanding that situation. There is ample evidence for flexibility and generalisability of the pragmatic schemas of street mathematics.
>
> *(ibid.* p. 147)

This is a surprising and telling claim. It argues that people engaged in street maths can use that activity *as a basis for expressing more general mathematical relationships*, and by implication, are able – at least to some extent – to reflect on what they do rather than simply engage in routinised practice. More broadly, the studies present powerful evidence against erecting an impermeable barrier between the informal and the formal, between the apparently unconscious and the reflective. They point to the existence of some linkage, some connection between what people do in their everyday practices, and the knowledge structures they can build while engaging in these practices – structures which, it seems, can be mobilised in broader contexts. Having an existence theorem is a start: but it does not, of itself, solve the problem of just how such linkages are constructed or can be facilitated. Fortunately, the Brazilian researchers throw some light on this question by focusing on the question of representations.

5.2. *The role of representations*

The key insight is that mathematical meanings constructed within a setting are inextricably interwoven with their representations. It was the distinction between oral (street) and written (school) representations which characterised the differences between mathematical activity in the street, in the market, in the drawing office, or in the classroom:

> It is possible that successful subjects not only have a schema of the situation in which the problem is being solved but can also keep it in mind more clearly as a consequence of the way the situation is represented. Thus, we need not only theoretical ideas that overcome the polarisation between general and particular knowledge but also ideas that bring to the fore the importance of forms of representation in thinking.
>
> (*ibid.* p. 144)

In coming to this conclusion, Nunes and her colleagues draw on the theoretical work of Vergnaud (1982) who argues that a mathematical concept involves a set of situations that give it meaning, a set of invariants constituted by the relationships essential to it and a set of symbols used in its representation. Thus structure, context *and* representation all comprise major pillars in the theory. Drawing on Vergnaud's insight, Nunes et al. show that simply looking at invariants identified through reflection on actions was not sufficient to characterise differences between practices – at least as far as *mathematical* modelling was concerned – because of the structuring role of the different types of representation adopted in problem solving.

Now this line of argument, insightful though it is, seems to treat as unproblematic the notion of representation. This, as any reader familiar with the literature will know, is nothing short of an intellectual minefield, and we have no intention of delving more than superficially into it here. Nevertheless, the issue of representation is important for what follows. The issue is, as usual, that of meaning. For Nunes and her colleagues, the differences in representation which accompany different settings are crucial in two ways. On the one hand, the differences facilitate flexibility and generalisability in different mathematical relationships, and allow attention to be focused on a range of metonymic (non-referential) relations. On the other, these very differences erect mental boundaries around mathematical ideas; the studies provide ample evidence of the 'disconnection between people's knowledge of street and school mathematics' (*ibid.* p. 147).

The key point is that representation in this context means oral or pencil-and-paper, representations which are global in the sense that they are drawn *on,* rather than locally evoked – drawn *in* – and constructed in response to the specificities of problem solving in time and space. In this scenario, representations influence what people do. But what of the reciprocal effect of people's behaviour on representations? What of the potential of representations which are constructed *in situ*? Can we explore the potential of child-constructed representations and how these representations and the child's conceptions mutually constitute each other dialectically?

This possibility has been a major concern of workers in the psycho-semiotic tradition. For example, Walkerdine (1988) and Solomon (1989) both point to the problematic relationship between signifier and signified. The question is not merely a matter of representation: 'Rather, the discourse itself is a point of production and creation' (Walkerdine, 1988, p. 203). From this perspective, the setting – and in particular, the representations available in it – are not simply constitutive of the individual's behaviour; the individual is simultaneously creating meanings within and for the setting.

Viewed from this perspective, arithmetic is not a simple matter of representing a certain structure or form from activity, but a much more complex process involving the individual's production of meaning in relation to social practices. A classic example is given by Walkerdine who analyses young children's ideas of 'more'. She shows how the practices in which they are engaged (cooking, playing a game etc.) shape the meanings for the word 'more', and that these meanings are more or less independent of the pedagogic practices which constitute school-arithmetic uses of the word. In the kitchen, for example:

> The dimension of *less* is simply not relevant and does not form part of that practice. The opposite of *more* in food regulatory practices is something like *no more, not as much,* and so on. *More or less* only form a contrastive oppositional pair with respect to certain practices and these practices are pedagogic.

> (*ibid.* p. 26).

This kind of analysis focuses attention on the role of representation as a central constitutive element of meaning-making. More fundamentally, it indicates how the representation structures *and is structured by* the context. Thinking of embedding tasks in 'context' (in ways beloved of mathematics text books) does *not* necessarily provide for learners any extra handle on the mathematical meanings; such a strategy might just as easily offer a complex and potentially confusing mix of readings which evokes more than one set of meanings, practices and understandings. These 'multiple significations' are, moreover, endemic to pedagogic practice: they cannot be wished away any more than we might be able to find a practice which is without the specificities of its own meanings and relations within its own discourse.

This perspective allows us to extend a little further the insights provided by Nunes and her colleagues. The point is that the power of the mathematical form is not that it suppresses meaning by not representing referents: it is that new meanings are created in the process (or, if they are not, the individual is unable to operate within the new discourse – a familiar enough situation for mathematics teachers!). It is not that meaning is necessarily replaced by 'syntactic rules' (although this, of course, *is* often the case), it is that one set of meanings is overlaid by another. The meanings of 'syntactic rules' are not necessarily surrogates for referential meaning; they can become another form of meaning, governed by the discourse of mathematical as opposed to everyday activity. The rules of discourse of fish-selling or basket-weaving have meaning only to the extent that the activities themselves are meaningful for the fisherman or weaver. In exactly the same way, the extent to which syntactic rules have meaning is the extent to which the discourse constituted by those rules evokes a set of meanings for the individual. The representational form within the discourse is important, and Nunes' work illustrates just how crucial it is. But written representations *can* be used – as mathematically-able adults know – as a way of structuring the discourse of a problem, and infusing it with meaning just as oral representations come with the connotations and denotations of the setting from which they are derived.

From this viewpoint, we can offer relatively straightforward explanations for findings like those in our N-task. The key lies in the ways in which the representational form (in this case programming code or pencil and paper) structured the expression of the children's proportional ideas. At the same time, by looking at the evolution of those ideas, we can begin to see how their development – from additive strategies to proportional reasoning – was structured by the representational form available. We might say that the program 'encapsulated' the logico-mathematical structure of proportionality: but in the light of the above, it seems more sensible to say that the program code shaped and was shaped by the children as they strove to infuse the idea of pro- portionality with meaning. The activities and their associated representational form became a resource for the expression of the relationships.

We will have more to say on this subject in later chapters, especially in Chapter 5. For the moment, we want to emphasise that the analysis we have outlined emphasises that learning mathematics is not simply a matter of decontextualising structure or form from activity in the physical world, but is a more complex process involving social practices, and the representations which accompany them. School mathematics is one such practice, but not the only one.

We are almost done with laying the foundations. But we have still to find a way out of the situationist cul-de-sac which simultaneously recognises the key role of context and representational form, yet preserves what is powerful about mathematical thought.

6. A WAY OUT OF THE CUL-DE-SAC?

One point which emerges from our analysis so far is that representations and the communicative devices with which they are intimately bound up, can no longer be regarded as neutral players in the process of making meaning. The question of language, and its relation to thought and activity, is therefore critical. Accordingly, we need to lay a further foundation stone which focuses on the relation between intellectual tools and mathematical thinking; to do

so, we consider the work of the Vygotskian school, whose contribution we
have not yet addressed.

6.1. *Language and tools*

Vygotsky's challenge was to understand how children become conscious subjects through the appropriation of forms of social activity that sustain symbol
systems, how knowledge evolves at the intersection of thought and language
(see, for example, Vygotsky, 1962). For Vygotsky, learning is mediated by
language, and signs are representational tools developed from social interactions. Vygotsky's work is of particular interest for those involved in the
study of mathematics education, since mathematics comprises a duality of
objects and relationships: there is a need to take into account how objects are
characterised and represented, the tools by which we act on them as well as
the language which describes how they interrelate. This duality is evident in
the origins of mathematics as a tool and as an object of study communicated
in particular linguistic forms.

Vygotsky suggests that the structure of mental activity is contingent upon
the symbol system available: 'According to Vygotsky, basic psychological
processes (abstraction, generalisation, inference) are universal and common
to all humankind; but their functional organisation (higher psychological
processes, in Vygotsky's terminology) will vary, depending on the nature of
the symbol systems available in different historical epochs and societies and
the activities in which the symbol systems are used.' (quoted in Nunes, et al.,
1993, p. 27). It therefore follows that in interpreting a response to any problem
situation, we have to take account of the activity, its goals and purposes, the
representations used in the modelling process and the symbol system with
which these representations are expressed.

But Vygotsky goes further. He insists that the approach to higher cognitive
function is through *decontextualisation* – sign-to-sign mediation (Wertsch,
1985). In other words, Vygotsky's criterion for higher cognitive function is
precisely what Nunes calls 'not representing referents', breaking with all
meanings derived from a referential domain. In doing so, he conclusively
allies himself with those who place thought and action in a hierarchical
relation, and – from an educational point of view, worse – break the connection
between them. Confrey puts it thus:

> ...in directly replacing the tools of labour with the psychological tools of
> signs, he (Vygotsky) virtually severed the connections between signs and
> the underlying tools, actions and operations that produce them and placed
> his emphasis solely on movement among the signs.
>
> (Confrey, 1995, p. 207)

Vygotsky's method was to focus on the meaning of the word. And by making
this choice, he necessarily privileges the formal, theoretical and mental over

the informal, practical and enactive. One consequence is that despite the rhetoric of cultural relativism in the Vygotskian school, mathematics appears as the epitome of decontextualisation, the pinnacle of abstraction. Bakhtin (alias Volosinov) makes the point forcibly:

> What interests the mathematically-minded rationalists is not the relationship of the sign to the actual reality it reflects nor the individual who is its originator, but the relationship of sign to sign within a closed system already accepted and authorised. In other words, they are interested only in the inner logic of the system of signs itself, taken, as algebra, completely independent of the ideological meanings that give the signs their content.
>
> (Volosinov, 1973, pp. 57–58)

The crucial point is this: the Vygotskian tradition points to mathematical discourse as a unique form, contrasting with all other sign systems. It usefully draws attention to the ways in which meaning is produced in terms of intra-mathematical relations, in sign-sign mediation, but it views this as the only mechanism for the production of mathematical meaning. There is no effective role either for other symbol systems, or interaction with objective social or physical reality. If this were true, we might at least go some way to explaining the difficulty with which so many are enculturated into mathematical discourse, but we would do so by erecting (or maintaining) a rigid barrier between social and practical activity on the one hand, and mathematical thought on the other. We will see that this isolationist view of the mathematical is unnecessary; mathematics can – in certain essential respects – be viewed as a discourse like any other.

6.2. Abstract versus concrete?

One way of gaining insight into the boundaries between mathematical and practical activities, is to study mathematics-in-use. For the moment, it will help to think of learning mathematics broadly as a modelling activity – whether building models of 'real' problems or of new mathematical theories – and in that case there is need to agree on the mathematical invariants, and to hypothesise the relationships between them. At one extreme, we could view this merely as a matter of dissociating them from the situation while naming and relating the mathematical invariants of the model. At the other extreme, we could recognise the complexity of any situation to be modelled and the myriad of potential meanings for any invariants according to their representations and their meanings.

To gain insight into this dichotomy, we need once again to introduce a social dimension into our investigation of meaning-creation. We will move well outside the domain of mathematics education, and incidentally, take a further opportunity to exploit the computer as a window on the problem. We draw on a fascinating study in which the computer was introduced to model forest valuations in Sweden (Göranzon, 1993). Göranzon describes the conflict between the central office staff including those responsible for the computer models, and the forest rangers who repeatedly stressed the importance of manual calculations if they were to maintain and exploit their professional expertise. The decontextualisation endemic to the modelling process ruptured the connections between meaning derived from practice and meaning derived from calculation:

> He [one of the Forest Rangers] talks about manual calculation as an imme-diate participatory experience in which one is actively involved at every stage. He chooses sensory and physical metaphors to describe the nature of this experience: the material was more alive, it was shaped in his hands, he saw if it was out of balance. He emphasised how his judgement emerged in the cause of the manual calculation and pointed out that one of the effects of computerisation had been to make it more difficult to uncover errors.

> (*ibid.* pp. 31–32.)

Göranzon's work underlines the dialectic between experiential, tacit knowl-edge and intellectual, 'scientific' knowledge. As tools are used (in this case arithmetical tools) they are adjusted in use so as to retain the sense of the practice – the professionals are able to use their judgement in setting up and interpreting the models. Their antipathy towards the computer model stems from the rupturing of the relationship between practice and theory, activity and mathematics.

One way of interpreting the rangers' views is that the computer models made it more difficult, perhaps impossible, to exploit the texture of their experience as a resource in their mathematical activity (this tells us some-thing about the particular models' rationales and implementation, rather than computer models *per se*).

The standard view of the modelling process is solely one of abstraction through decontextualisation: there is little room for a social dimension, or for considering the relationship between tools and model. The assumption is that modelling necessarily involves ascent to a higher realm of cognitive activity: a presupposition that there exists a hierarchical relationship between mathematical and social activity.

Yet Göranzon's example suggests that knowledge is constructed in the course of clusters of interrelated activities. Perhaps the abstraction process might be thought of not so much as a step upwards, but rather as an inter-twining of theories, experiences and previously disconnected fragments of knowledge including the mathematical. Perhaps it is possible to think of model building as a dialectical relation between the practical and theoretical, rather than a simple ascent from one to the other?

This possibility resonates with a number of studies which have recognised the fundamentally fragmented nature of knowledge building (see, particularly, diSessa, 1988a; Papert, 1980). Similarly, the work of Cheng and Holyoak

(1985), though from a different and more psychological perspective, points in a similar direction. While seeking to understand the way schemas of if-then reasoning were developed, they posited the inadequacy of the specific-experience view and proposed the notion of 'pragmatic reasoning schema' which are guided as much by goals and purposes as the logic of the argument and are 'primarily products of induction from recurring experiences with classes of goal-related situations' (*ibid.* p. 414). The claim is that '...people typically reason using abstract knowledge structures organised pragmatically rather than in terms of purely syntactical rules of the sort that comprise standard logic' (Cheng, Holyoak, Nesbitt, & Oliver, 1986, p. 314).

Thus, a pragmatic schema is operationalised over a variety of situations that share some common properties. As we have seen, Nunes *et al.* apply this idea to street mathematics: we see how people using money could carry out transactions which obeyed mathematical rules but were not applied in a decontextualised way (for example, using the associative law of addition explicitly). Street mathematicians reason on the basis of social as well as logical rules. In this scenario, models are constructed and validated on the basis of economic, political and social rules alongside those of mathematics. Like Göranzon's rangers, there is an interplay between model building – mathematical activity – and lived-in experience, rather than the dominance of one over the other.

At this point, we introduce explicitly a dichotomy which permeates the literature on mathematical meaning-making but which we have so far only hinted at: the dichotomy between concrete and abstract. A standard description of the difference between thinking in lived-in experience and mathematical thinking is that the former is concrete, the latter abstract. There is no doubt which way the hierarchy sits: the history of Western thought has privileged the latter at the expense of the former. Abstract is general, decontextualised, intellectually demanding; concrete is particular, context-bound, intellectually trivial.

The difficulty with the hierarchical view is that it implies that persons acting in and on their environment are 'only doing' – that practice is completely divorced from intellectual activity. Yet we have now become familiar with cases – Nunes' fishermen, Göranzon's rangers, and, indeed, our N-task children – in which the picture is far more complex than this simple hierarchy might suggest.[18]

We can make sense of the modelling process without setting up abstract and concrete as bipolar opposites. Rather than reorganisation and decontextualisation there can be closer interaction between the model and what is modelled, a recognition of the multiple generalities within different particularities. This does not have to be a simple carrying over of meanings derived from 'real' referents: on the contrary, the examples above show how these meanings are reshaped in the interplay between 'abstract' and 'concrete' activities. We have an image of mutually constitutive meaning-making in which both kinds of activities act as resources *for each other*.

Perhaps the most clearly elaborated vein of work seeking to construct an alternative view of the relation between abstract and concrete has emanated from Papert and his colleagues at the Massachusetts Institute of Technology. This work was derived from a critique of Piaget's one-way view of cognitive development, in which children become progressively detached from the world of concrete objects and local contingencies, and slowly but surely ascend to the level of abstract thought. For example, Turkle and Papert (1990) argue for the location of intellectual development within rather than beyond the concrete. This blurring of concrete and abstract is nicely illustrated within computational settings, which, they propose, afford the learner something in the domain of formal systems to which they can relate informally. In a similar vein, Ackermann (1991) provides a fascinating re-analysis of Piaget's water-level experiments in which she emphasises the coordination of local knowledge in building abstractions, and argues that the induction of general laws flows from the regularities children discover in specific contexts *while at the same time* applying general rules in order to make sense of local situations. For Ackermann, abstract and concrete are two sides of the same coin – as with situated and unsituated, they are dialectically interrelated rather than opposed.

So far so good. But there is a world of difference between water-level experiments and mathematics. Wilensky (1991) develops this theme, taking issue with what he terms the standard view of the concrete and points out that:

> objects... which are particulars in one ontology can be generalisations in another. Indeed for any concrete particular that we choose, there is a world view from which this particular looks like a generalisation.

> (*ibid.* p. 196)

Following Piaget, Wilensky's focus is on the interaction of knowledge (objects and relations) and the individual: 'concreteness is not a property of an object but rather a *property of a person's relationship to an object*' (*ibid.* p. 198; emphasis in original). This is a powerful way of thinking about concreteness which has surfaced in different forms within the mathematics education community. For example, Bishop (1988) has argued that '...an idea is meaningful to someone to the extent that it makes connections with other ideas known to that person' (p. 80). Wilensky takes the argument to its conclusion, and argues that if concreteness is a relational property then: 'The more connections we make between an object and other objects, the more concrete it becomes for us' (p. 198). To a topologist, a four-dimensional manifold is as concrete as a potato.

Meanings are created by experiences – experience can evolve from inter-action between mental states as much as between a mental state and the physical world. The fragility of meanings in mathematical worlds stems from unfamiliarity, from the difficulties of building experiential domains within them. A (abstract algebraic) ring is rich in meaning, so much so that there is at least one example (the integers) of a ring known quite intimately to almost every child. There *is* a difference between a ring and the friendly integers, but it is a difference of quantity not quality: it is determined by how rich the set of examples is, how connected the examples are to each other, how familiar the common properties are, the richness of the images and metaphors built and with what ease they can be mentally turned around. As Wilensky puts it:

> This view will lead us to allow objects not mediated by the senses, objects which are usually considered abstract - such as mathematical objects - to be concrete; provided that we have multiple models of engagement with them and a sufficiently rich collection of models to represent them.

(ibid. pp. 198–9)

Thus the hierarchy begins to look a little shaky. The challenge for the individual learner becomes one of constructing multi-faceted connections between activities and experiences that are 'in some way' similar. Abstraction becomes a problem of how to add new friends and relations, not to ascend to unattainable heights. And from this perspective flows the educational corollary that it might be possible to design educational environments in which the process of abstracting becomes part of the lived-in culture of experience.

6.3. *Returning to the vignette*

Before we try to cement together the foundation stones of this chapter, we briefly pause to reconsider our N-task vignette. A key observation was the identification of strategies and developmental paths among the task solutions which were heavily contingent upon the modality of interaction. From the vantage point we have established, we can begin to reconceptualise the data. The letter Z was planned spontaneously at the computer; the N also involved interludes using pencil and paper. In the N task, directing attention to the computer at key points in the problem-solving process, shifted paper-and-pencil strategies by highlighting the proportional relationship through the need to express it as a program. The 'addition with adjusted increments' strategy was 'transformed' by the switch to the computer, by the need to express it in Logo: it was expressed as adding a fraction of : S to : S – recall how David focused on the relationship between the sides of the N and reified the scale factor which he could not construct but knew 'existed' by naming it FORMULA. This expression of generality emerged as an integral component of David's construction of the relationships; somehow – in ways we will clarify in Chapter 5 – he mobilised the resources of the setting in order to simultaneously *abstract* and *give meaning to* the objects he was attempting to construct.

Something rather different happened when Paul and Noel built their Z. The children built a strong sense of connection between the new object (in particular the diagonal), its constituent parts and how it related to the set of other objects of the same form. They had constructed a meaning for the idea of variable, 'substituting values'. What happened next could hardly be characterised as abstraction in the traditional sense, a letting go of referential meaning. On the contrary, the transition to 'multiplying by :S', with its explicit focus on the relationship, was a matter of creating a new meaning, making a new connection through the medium of the code in which they expressed their problem-solution.

7. RETHINKING ABSTRACTION

We have arrived at a point where the nature of the abstraction process itself has emerged as crucial in understanding how mathematical meanings are constructed; it will form a recurrent theme in the structure we build on our foundations.

Our first foundation stone is our critical stance towards a view of mathematical meaning which hinges on the reification of lower level conceptions to pinnacles of the abstract. Our argument is that if we restrict our terms of reference simply to the interaction of epistemology and psychology – ignoring the social dimension – then it is *inevitable* that mathematical learning will be perceived as the acquisition of context-independent knowledge within a hierarchical framework. Starting from a position of epistemology/psychology locks research and its findings into a tautological loop. Moreover we do not see it as accidental that these hierarchies conveniently mirror the various sequences of lessons in the school curriculum and bolster the assumption that mathematics is only appropriate for a small elite.

Of course multiple hierarchies do exist on a local level – clearly there are interconnections, local chains of definitions and argument within and across a tapestry of concepts making up the mathematical domain. But this tells us little about the global nature of the subject, still less about the competencies of children.

Our second foundation stone rests on our characterisation (following Wilensky and others) of abstraction as a process of connection rather than ascension. Paradoxically then, we *do* regard abstraction as a central facet of the mathematical enterprise, and – by implication – of the business of learning mathematics. Here we might differ somewhat with others whose perspective we otherwise broadly share. For example, Confrey (1993) develops an elegant way of thinking about learning in which tools mediate both 'grounded activity' and 'systematic enquiry'. By grounded activity, Confrey is flagging her view that mathematics evolves from actions upon concrete objects, seeking to distance herself from the negative pedagogical connotations of abstraction.[19] Like us, Confrey is opposed to the view of increasing abstraction as necessarily emphasising detachment and decontextualisation. Unlike us, she is prepared to relinquish abstraction as a goal in favour of systematic enquiry. But in our view, abstracting in all its diverse guises is more than orderly investigation: we see no reason to admit that abstraction necessarily involves loss of meaning – quite the contrary. There is more than one kind of abstraction; this is a crucial point we shall elaborate in what follows. If we can find new ways to abstract which create new meanings, we can expand the scope of what it means to do mathematics without sacrificing the characteristics which give mathematics its power.

Our preference is for an image which emphasises actions on mathematical objects (like telling the Logo turtle to go 'FORWARD' or 'BACKWARD') and it is these types of actions which can be closely connected to abstracting. Meaning can be maintained by involvement in the process of acting and abstracting, building new connections whilst consolidating old ones. This is not a novel idea:

> ...perhaps this sense of being out of contact (referring to a common response to abstraction) arises because there has been little or no participation in the process of abstraction, in the movement of drawing away. Perhaps all the students are aware of is the *having been drawn away* rather than the *drawing* itself.
>
> (Mason, 1989, p. 2: emphasis in original)

If the preservation of meanings is paramount, we must focus more on continuity than difference, and in the process, re-appropriate the words 'concrete' and 'abstract', breaking the latter free from its dehumanising connotations. We have begun to point to the kinds of environments which have the properties we seek: those in which mathematical meanings can be constructed from an intersection of resources mobilised by actions, in which mathematical ideas become connected with existing understandings and activities. We have pointed to the evidence that suggests that the situated, the activity-based, the experiential can contain within it the seeds of something more general. The corollary is that we need to focus on the issue of representation of mathematical objects and how these are expressed; and more fundamentally, how the resources of a setting (be it supermarket shopping or Logo programming) mediate that expression.

This last point leads to our third foundation stone. The situationists want to conceptualise thought as structured by setting, shaped by the representational tools available within it. We would like to re-establish a focus on the reverse process: on the ways in which setting acts back on thinking. Tools are not passive (see Kaput, (1986) for a discussion of how tools can open new representational windows), and we would like to understand how the resources within a mathematical environment are structured by activity within it. Here, it seems, we are on common ground with the reification theorists. For example, Sfard and Linchevski state:

... it is fairly possible that the massive use of computer graphics in teaching functions will reverse the 'natural' order of learning so that the structural approach to algebra will become accessible even to young children.

(Sfard & Linchevski, 1994, p. 224)

Where does this leave the 'natural propensity' for an operational approach? If the relationship between operational and structural understandings is contingent upon tools, might it not similarly depend on context, specificities of task, and mathematical domain? Sfard and Linchevski's speculation concerning the computer is apt: and it illustrates two general points. First, if we want to study the trajectory of mathematical learning, we need to focus on tools and settings as well as on psychology and epistemology. Second, the computer focuses our attention on the relationship of learning to context, on the ways in which the understanding of mathematical ideas is mediated by the tools available for its expression.

It is for these reasons that we have chosen computational environments as sites in which to study the dialectic between activity and setting, between informal and formal thinking. We have glimpsed how the computer can, in ways we have yet to elaborate, combine elements of the 'concrete' with the 'abstract', the intuitive with the rigorous, the particular with the general. At the same time, it affords the learner a support for thinking, a screen on which to construct and reconstruct emergent ideas, and on which we, as observers, can catch sight of that construction process as it takes shape.

It should already be clear that the specificities of a computational environment are paramount – we had better place on record the manifest absurdity of attributing any property to 'the computer' in general, despite the common abuse of language to which we shall sometimes default. In the following chapter, therefore, we will set out the rationale for our preference for studying mathematical thinking in computational settings, and elaborate the characteristics of environments which provide a clear window onto the construction of mathematical meanings.

NOTES

1. This language has evolved and transformed throughout the history of mathematics.
2. These were: measurement, number operations, place value and decimals, fractions, positive and negative numbers, ratio and proportion, algebra, graphs, reflections and rotations, vectors and matrices.
3. We note that the legacy of this early research has provided the theoretical legitimation for recent educational policy in the U.K. and elsewhere.
4. The fundamental flaw in this argument is discussed in Dowling and Noss (1990) and briefly illustrated in Chapter 7.
5. This movement emanated from the U.S. and never really travelled across the Atlantic.

6. For example, how an individual proves a mathematical conjecture depends on knowledge and experience, the techniques and resources available and what will be accepted as a proof within the community.

7. We note in passing the complexity of this issue – the setting constraint of multiplication can be discarded in the following example while still yielding an answer which is mathematically correct:
$$dy/dx * dx/dt = dy/dt$$

8. One view, due to Walkerdine (1988) is that there are chains of signification associated with any mathematical idea so that an individual might adopt multiple and even contradictory positions in the face of any problem situation, and that these would be triggered independently of each other according to the influence of different settings. We will discuss Walkerdine's central idea later in the chapter.

9. There are many branches of constructivism: classical, social, radical. Here we largely stay with classical constructivism rooted in Piagetian theories.

10. It is questionable how far these facets can be derived from the philosophy and epistemology of constructivism *per se*.

11. Didactic: our thesaurus says terse, obvious, concise. This confusion is evident only in anglo-saxon cultures, not to the languages of mainland Europe, where the term is in general use.

12. The last phase in the cycle is that of *validation*.

13. The term 'advanced' is used in contradistinction to the emphasis in the mathematics education community on early school mathematics. It ranges from late school to early University mathematics, and subsumes for example, elementary calculus as well as more abstract mathematics such as group theory and analysis.

14. The following section is based on a report of the study, published in the Journal of Mathematical Behaviour (Hoyles & Noss, 1989).

15. We omitted any category involving formal rules: for example the 'rule of three' or a similar algebraic relationship. Our children had not been introduced to this or anything like it in their mathematics lessons.

16. Gender bias in tests is one well-researched area related to these discussions.

17. We are hinting here at the way we are going to get out of this cul-de-sac, which will involve the redefinition of notion of referent to include intra-mathematical relationships. As usual, the issue is one of meaning, not 'reality'.

18. As an aside, it is worth recalling the commonplace observation – surprising when it first emerged from the Artificial Intelligence community – that it is much simpler to build a computational system which plays chess, than to construct a program which can recognise a face or open a door.

19. Like us, Confrey is deeply influenced by the notion of computer as tool, and argues : '...these machines (computers) are tools like hammers and scissors, but they are also displays of symbols systems, and thus of psychological signs. Hence they need to be viewed as mediators of social-cultural participation' (*ibid.* p. 53).

CHAPTER 3

TOOLS AND TECHNOLOGIES

1. COMPUTERS AND EDUCATIONAL CULTURES

It has become part of accepted educational wisdom that the computer by itself cannot fundamentally change either what is learned or how, and that issues of learning and teaching are dependent on more than the simple presence of the computer in the learning situation. This does not prevent a substantial majority of educational research from treating the computer as a thing-to-be-researched, or policy makers from viewing the mere provision of hardware as a determinant of educational change.

There is a reverse tendency which has become evident in some circles, a tendency which, while very properly refusing to be carried along by the wave of technocratic enthusiasm, fails even to consider the possibility that the computer might fundamentally influence educational practice in anything but a negative direction (for a bizarre example of this position, see Robins and Webster (1989), who accuse Papert of a vision for education which 'is an echo of military systems discourse' *ibid.* p. 259). Critiques of this kind are right to draw attention to the roles to which computers are put and the rationales for their existence, but they throw out the educational baby with the sociological bath water. Recognising that the technology is not, can never be, an independent agent misses the uniqueness of what the computer has to offer. It ignores the dialectic between how cultures structure technology and how technology can shape the culture which gave birth to it.

This is a convenient place to pause and examine our use of the term *dialectic*, which we employed without elaboration in the previous chapter. A common abuse of language has lent the word an impotent ring, in which it has come to mean merely 'opposed' or 'two-sided'. There is more to the term than this, and it is worth a digression to outline our interpretation from a brief study of its etymology. Friederich Engels, the collaborator of Karl Marx, traces the origins of the word from the Greek philosophers, through 'brilliant exponents' of the dialectical method such as Descartes and Spinoza, and culminating in the idealistic dialectic of the German philosopher, Hegel. For Engels, the key idea of dialectics is the move beyond the crucial yet restricted recognition of the complexity of social and material reality, and in particular, beyond the metaphysical view in which '...things and their mental reflexes ideas, are isolated, are to be considered one after the other and apart from each other, are objects of investigation fixed, rigid, given once for all' (Engels, 1968, p. 406). In this view, 'positive and negative absolutely exclude one another; cause and effect stand in a rigid antithesis one to the other' (*ibid.* p. 406).

In contrast, the dialectical view sees the two poles of an antithesis as inseparable at the same time as opposed:

> And we find, in like manner, that cause and effect are conceptions which only hold good in their application to individual cases; but as soon as we consider the individual cases in their general connection with the universe as a whole, they run into each other, and they become confounded when we contemplate that universal action and reaction in which causes and effects are eternally changing places, so that what is effect here and now will be cause there and then, and vice versa.
>
> (*ibid.* p. 407)

Not surprisingly, our vision is somewhat less restricted than that of Engels', whose project was nothing less than to understand, and change, the economic and social structures of the capitalist world. Nevertheless, there is much we can borrow from his elaboration of the dialectical method, not least for the methodologies which are appropriate to explore the relationship between technology and education. For the great power of the Hegelian system of thought was that: 'for the first time the whole world, natural, historical, intellectual, is represented as a process, i.e., in constant motion, change, transformation, development...' (*ibid.* p. 408).

Our concern is to understand – and to change – the 'whole world' of mathematical education. And it is to this end that we make our emphasis in this chapter that of the technology itself; if this emphasis is thought to be one-sided, it is an imbalance which we will amply counteract in the remainder of this book.

In fact, the mistake of believing that it is culture which *determines* technology is a mistake which, according to Engels, was characteristic of the Hegelian dialectic – a mistake which stems from the fact that Hegel's dialectic was essentially *idealist*. For Hegel, 'the thoughts within his brain were not the more or less abstract pictures of actual things and processes, but, conversely, things and their evolution were only the realised pictures of the 'Idea', existing somewhere from eternity before the world was' (*ibid.* p. 408).

Our vision of mathematical learning includes the computer as an important part of the material realities of the learner and teacher, with the potential to catalyse fundamental changes in them. That some are reluctant, on the basis of the ubiquity of electronic drill thinly disguised as 'games' or 'adventures', to engage with the computer as a serious element of the educational enterprise or to recognise interaction with such a system as in any way comparable to a linguistic tool, is understandable but mistaken. Our first concern then, is to clear the decks for consideration of computational systems that *can* lay claim to be taken seriously as putative contributors to the learning of mathematics.

2. A PRELIMINARY CASE FOR PROGRAMMING

There is a major fault line which runs through computer software designed for learning. On one side lies software which is designed to deliver existing mathematics curricula, to repackage mathematical knowledge and present it in acceptably wrapped form for educational consumption. On the other, lie computational applications which point outward towards new, more learnable, mathematics; towards a redefinition of what school mathematics might become and who might become involved in it.

Our interest centres firmly on the latter vision, and for two reasons. The first is pedagogical. Our preference for excluding software aimed at teaching a given curriculum centres around the word 'teach'. There may be room for such software: one way, presumably, to teach the solution of a quadratic equation is to provide learners with, say, an animated screen which displays the formula. But we find such software fundamentally uninteresting from a pedagogic point of view, in that it offers nothing more – perhaps substantially less – than a human tutor. The last decade has seen the mathematics education community struggling to outlive the legacy of the early days of computers in schools, which were characterised by countless programs thrown together by well-meaning amateurs, claiming to teach this or that mathematical topic. These programs offered little if anything more than textbooks – at considerably greater cost and inconvenience – and did much to turn a generation of mathematics teachers against computers altogether.

This relates to our second, more fundamental objection. Software which fails to provide the learner with a means of expressing mathematical ideas also fails to open any windows onto the processes of mathematical learning. A student working with even the very best simulation, is intent on grasping what the simulation is demonstrating, rather than attempting to articulate the relationships involved. It is this *articulation* which offers some purchase on what the learner is thinking, and it is in the process of articulation that a learner can create mathematics and simultaneously reveal this act of creation to an observer.

There was a time when one could suggest that such articulation was only available in a very few systems: programming languages, the professional 'tools' (like spreadsheets) and one or two others. Now, there is a small but significant attempt to address with computational systems, problems of mathematical teaching and learning from perspectives other than crude behaviourism and curriculum delivery. It is not our purpose to catalogue worthy programs, not least because such a catalogue would inevitably be out of date by the time this book appears. Instead, we will continue to use Logo as a paradigmatic case for the approach we favour. We will, however, also present examples of other systems, most notably dynamic geometry systems,[1] which claim to provide an alternative to the programming route: contrasting dynamic geometry tools with Logo will also afford valuable insight into the mediating role of the computer.

There is a long pedigree to the idea that writing a computer program provides a broad canvas on which the learner can sketch half-understood ideas, and assemble on the screen a semi-concrete image of the mathematical structures he or she is building intellectually. Early research produced some robust indications that the expression of mathematical ideas in the form of computer programs suggested a promising line of enquiry. This pioneering work included programming both in BASIC (see, e.g., Hatfield & Kieren, 1972) and Logo (an early report is given in Feurzeig, Papert, Bloom, Grant, & Solomon, 1969).

Since then, further investigation has largely fallen to those working within the Logo community, although an alternative approach, more agnostic on choice of language, has continued under the banner of 'algorithmics' (see, for example, Johnson, 1991). The comparative isolation of the programming community within mathematics education, has largely been due to the notion of programming held within the wider educational culture (this point is discussed in detail in Chapter 7). A programming language is, after all, for writing programs: lines of text are entered, output is (hopefully) obtained. Compared with the visually attractive screens of multimedia systems, or even the direct manipulation interfaces characteristic of the dynamic geometry tools, programming can seem dull, arcane and unnecessarily complex. Yet recent developments in the nature of programming have changed even this situation, perhaps not fundamentally, but at least enough to challenge the most obvious cultural stereotypes of the programming idea.

First, the notion of programming has fundamentally shifted to include visual and iconic rather than purely textual means of combining elements (as, for example, in *Function Machines* described in Cuoco, 1995, and new versions of Logo which incorporate screen gadgets, such as sliders and buttons). This is only an indication of what is to come. The computational medium *Boxer*, in some ways an extension of Logo, offers much more: it is a system whose different functionalities are glued together by programming (in a language which is in the Lisp/Logo tradition) and in which every aspect can be directly perused and manipulated by pointing and executing. Boxer offers the learner a concrete visual model on the screen of what is happening in the machine — rather than the learner having to create his or her own mental model independently. It aims to provide the user with much greater expressive power:

> We are trying to produce a prototype of a system that extends with computational capabilities the role now played in our culture by written text. It should be a system that is used by very many people in all sorts of different ways, from the equivalent of notes in the margin, doodles and grocery lists – all the way to novels and productions that show the special genius of the author, or the concerted effort of a large and well-endowed group. In

a nutshell, we wish to change the common infrastructure of knowledge presentation, manipulation and development.

(diSessa, 1988b, p. 3)

Second, powerful microworlds have been designed which are tuned to a knowledge domain while retaining the full functionality of the Logo programming language. For example Sendov and Sendova (1993) have built a microworld for Euclidean Geometry, *Geomland*, which offers some of the same functionality of a dynamic geometry system while being written in Logo. In exchange for direct mouse manipulation, Sendov offers a text-based interface to explore geometrical objects and relationships. The advantages and disadvantages of direct manipulation and text-based interfaces for the learning of mathematics have been the subject of heated debate (see, for example, diSessa, Hoyles, & Noss, 1995): proponents of the former point to a sense of engagement developed with screen objects; advocates of the latter stress the language of description and communication which programming affords.

New developments are beginning to break down the antithesis between programmability and direct manipulation, and point towards new systems which exploit the strengths of both. The new situation has been described eloquently by Eisenberg (1995) who has argued for what he terms *programmable applications*:

> Programmable applications ... are software systems that integrate the best features of two important paradigms of software design – namely, direct manipulation interfaces and interactive programming environments. The former paradigm - popularly associated with menus, palettes, icon-based interaction techniques and so forth – stresses values of learnability, explorability, and aesthetic appeal; the latter, by providing a rich linguistic medium in which users can develop their own domain-oriented 'vocabularies', stresses values of extensibility and expressive range.

(Eisenberg, 1995, p. 181)

In both the Sendovs and Eisenberg's vision (the latter's choice of programming language is Scheme – a dialect of Lisp – not Logo[2]), the central idea is to open up to the learner the expressive power of programming as a means to navigate and reconstruct a domain:

> Rather than viewing programming as beyond the capabilities of students, or as antithetical to the design of learnable software, the programmable-application designer instead creates software that allows students to work with both a powerful (and ideally, learnable) interface and a powerful (and ideally, learnable) programming language. If properly designed, the complexity of such an application should grow with use, as students develop larger and more advanced programs within the application; that is, the complexity of the application should derive from the student's ideas, and

not from the sprawling size of the interface. Moreover, by representing their ideas in a progressively elaborated linguistic medium, students are given an opportunity to grow with the application over time–perhaps, over a lifetime; the transition between 'educational' and 'professional' activity is seamless.

(ibid. p. 181)

It is clear that Eisenberg's vision includes programming as an activity for everyone, rather than code produced by professionals for others to consume. In line with this vision, it is now common for systems of all kinds – from word processors to spreadsheets – to include elements of the programming idea. Even Cabri Geometry, which is highly tuned to a specific knowledge domain, allows the user to peruse histories of commands and construct 'macros' whose instructions can be re-used in different constructions, and may soon be editable just like a program.

We regard Eisenberg's vision as worth pursuing. It is no accident, therefore, that the examples in the rest of this chapter focus on programming in one form or another; in all cases programming is represented as a tool for expression and articulation. Yet how crucial is this for learning mathematics? Most importantly of all, how far can a focus on programming in particular serve as a window on mathematical expression in general?

2.1. Computers and conviviality

We have suggested that programming in its widest sense might be thought of as a tool for expressing and articulating ideas. Illich has called such tools *convivial:* conviviality is a question of *meaning.*

> To the degree that he masters his tools, he can invest the world with his meaning; to the degree that he is mastered by his tools, the shape of the tool determines his own self-image. Convivial tools are those which give each person who uses them the greatest opportunity to enrich the environment with the fruits of his or her vision.

(Illich, 1973, p. 21)

The extent to which a tool may be seen as convivial is the extent to which the use of the tool creates meanings for its user, catalyses intellectual experience and growth. We do not claim that *the computer* is convivial, any more than Illich claims that a tool is or is not convivial; it is not difficult to think of examples of computers in mathematical learning situations which are anything but convivial. On the other hand, it is our view that the computer is a tool which can be used to enrich the social and psychological space of the individual with the fruits of his of her (mathematical) vision.

Here we can establish a natural link with programming. Illich says:

> Tools foster conviviality to the extent to which they can be used, by any-body, as often or as seldom as desired, for the accomplishment of a purpose chosen by the user. ...*They allow the user to express his meaning in action.*
>
> (Illich, 1973, pp. 22–23, emphasis added)

Meaning expressed in action. Here is the heart of the matter as we left it in Chapter 2. We have seen that the computer is a rather special tool in which action involves the formal use of language, and where the usual polarities — meaning and precision, intuition and formalism, conviviality and rigour — do not hold. For mathematical learning this is a critical facet of the computer's role. As Papert points out:

> Most children never see the point of the formal use of language. They certainly never have the experience of making their own formalism adapted to a particular task. Yet anyone who works with computers does this all the time.
>
> (Papert, 1975, p. 220)

A central part of subsequent chapters will be aimed at addressing the claim that in school mathematics the computer can be a convivial tool. How does the computer allow the user to express mathematical meaning in action? What does Papert mean by 'anyone who works with computers', and what precisely is it that they 'do all the time'? What are the implications and limitations of the metaphor of the computer as tool?

A hammer is a tool.[3] It performs, more or less adequately the task which it is designed to perform. It can be used creatively (hammers can be used to prop open doors as well as hammer in nails); and certainly there is knowledge and understanding which is mobilised when using it. A piano is also a tool: a tool for making music rather than for hammering in nails, but a tool nonethe-less. Yet this designation does not sit easily. The uneasiness stems from our *relationship* with the tool; not the undeniable fact of the tool's usefulness, but the way it is used, the way we *feel* about it. The piano feels like more than a tool, it feels convivial. To be sure, many people are as indifferent to pianos as they are to hammers; and conversely, just such a convivial relationship is surely possible between a woodworker and his/her hammer, or lathe, or screwdrivers. The essential difference is the way in which the tool enters the personal and cultural space of its user.

Once we are clear about this feature of tool-use, it becomes easier to clarify how we think about computers. We are typing this book on computers which are convivial tools *for us*. For most people involved in computer use, the computer is far from convivial. Far from 'investing the world with his vision', the computer user is 'mastered by his tools'. There are two key observations. First, tools are cultural objects. Tools are not passive, they are active elements of the culture into which they are inserted. Second, the extent to which a tool is convivial is determined by the relationship of user to tool, not by any ontological characteristic of the tool itself.

For a tool to enter into a relationship with its user, it must afford the user expressive power: the user must be capable of expressing thoughts and feelings with it. It is not enough for the tool to merely 'be there' (like, say, an aeroplane is for a passenger), it must enter into the user's thoughts, actions and language (like, say, an aeroplane is for a pilot).[4] Expressive power opens windows for the learner, it affords a way to construct meanings. Just as importantly, it justifies our focus on programming languages in what follows. If learners can express what they mean, we might get a glimpse of these meanings ourselves.

3. THE DEVELOPMENT OF A PROGRAMMING CULTURE

Two general issues emerged from early research in the field of programming with Logo. In the first place, it became evident that programming was 'difficult' and 'time-consuming'. Neither of these findings were particularly surprising, except for those caught up in the euphoria of the time which tended to misrepresent Logo as being 'easy' and 'quick' to learn. In fact, research on this issue added little to common sense. If students need to acquire expressive power in a programming language, this surely must require time – perhaps as much as learning a foreign natural language in school (although as Papert points out, little apparent effort is needed for this if one lives in the right country for a short period), but a substantial commitment nonetheless.

This 'difficulty' could be looked at in a different way. Suppose the community of educators decided it was worth this commitment. Suppose educators were prepared for children to learn, at an appropriately early age, to program in Logo alongside the other 'basic' priorities of the school curriculum – like 'numeracy' or 'literacy' – in order that Logo became just another language in which ideas could be expressed.[5] What then? One corollary would be that the mathematical curriculum would need to change, and change radically. Perhaps certain pieces of mathematics would lose their importance? Perhaps the sequencing of topics would need urgent revision, as some mathematical ideas could more easily be learned with programming at hand? Perhaps the time spent on learning to program could be more than made up by the time saved on speeding up the learning of standard mathematical ideas, or on disposing altogether with obsolete ones?

3.1. *The development of a literature*

It is one thing to recognise that learning to program is comparable with the effort of learning to read. It is another to ensure that there is a literature worth reading. The development of a viable *LogoMathematical* literature at the interface of Logo and mathematics, at least one which is accessible for both young and experienced learners, has been rather slow.[6]

When Papert wrote on the dust jacket of Abelson and diSessa's *Turtle Geometry* (1980) that it was the 'first textbook for the mathematics education of the future', he was pointing towards the construction of a culture which would develop new kinds of learnable mathematics through programming. In reality, there was an appreciable time lag between its publication (the same year as *Mindstorms*) and its worthy successors. As a result, a substantial volume of research was undertaken seeking 'effects', before it was clear what might be responsible for any effects which might be found: much as wondering what the effects of learning to read might be, before there are any books (we shall discuss one such study in detail in Chapter 7). Fifteen years later, there are now important landmarks in the construction of a new mathematical culture which rank with Abelson and diSessa's achievement. These include Al Cuoco's *Investigations in Algebra*, James Clayson's *Visual Modelling with Logo*, Wally Feurzeig and Paul Goldenberg's *Modelling Language with Logo* and Philip Lewis' *Approaching Precalculus Mathematics Discretely*. These books, together with countless nuggets of shared experiences of teachers and academics (in, for example, professional journals such as *Micromath* and *Logo Exchange*) have helped to create an outline of new mathematical activities and experiences for learners. The appearance of this literature – as yet gradual and not highly disseminated – can hardly be overestimated in the drive for cultural change.

3.2. *The development of a research community*

Alongside this creative enterprise, there emerged a small community of Logo-Mathematicians, who have addressed seriously a range of issues associated with the learning of mathematics with Logo. Much, though not all, of this research enterprise emanated from an international group of scholars who came together in the early nineteen-eighties. Their common interests lay at the intersection of Logo and mathematics, and in their growing realisation that in the heat of the Logophile eighties, the mathematical grounding of the Logo idea was in danger of being lost. For the most part, they attempted to maintain the spirit of Papert's vision elaborated in *Mindstorms* but with the specific intention of analysing how school mathematics and its transaction might be shaped by Logo.

This group – which became known as *Logo and Mathematics Education* (LME) – held regular conferences starting with the first, in London (Hoyles & Noss, 1985). It evolved a commitment to new kinds of mathematical learning within the school curriculum (for example fractal geometry); the revitalisation of previously esoteric mathematics by exploration with programming; the development of new theorems and processes of validation; and finally and more broadly, the forging of connections with 'kid culture' derived from computer use in the home and the video arcades. At the same time, many within the group remained committed to designing environments, *microworlds*[7] which would operationalise the vision of computer-based change both within school and without.

Among the LME group, and the group loosely affiliated to Papert and his co-workers at MIT, there was thus a twofold sense of purpose. First, to design playful, informal, and mathematical worlds in which learners could explore and express mathematical ideas and relationships in diverse ways. Such worlds inevitably excluded much that characterised conventional mathematical learning: repeating routines, parroting algorithms or guessing the teacher's agenda. The second common purpose was to generate worthwhile research questions and elaborate appropriate methodologies with which to investigate them. This resulted in a research literature which has addressed the issue of just what children can do mathematically with Logo, as well as what they might learn; investigations have included longitudinal studies of learners' LogoMathematical work, focused studies on algebraic and geometric thinking, advanced mathematical topics such as group theory, mathematical modelling, and reconceptualisations of children's problem-solving strategies. Many, though by no means all of these studies have been compiled in a volume we edited not long ago (Hoyles & Noss, 1992).

This dialectic between the design of learning environments and the research effort to describe student's learning within them has become a crucial part of the LogoMathematical research enterprise, and lies at the heart of Papert's 'Constructionist' paradigm. Constructionism – the idea that learners build knowledge structures particularly well in situations where they are engaged in constructing public entities – has provided the framework for the strand of Logo-based research centred around MIT, and has given rise to detailed investigation which ranges widely over topics as diverse as knot theory, learning styles, and musical composition. This wide sweep of constructionist research has, for the most part, adopted a less recognisably mathematical focus than that of the LME group, but one which has generated a substantial and impressive corpus of findings (many of these studies are collected in Harel & Papert, 1991).

3.3. *Summing up the case for programming*

Our focus on programming stems from our interest in the rigorous articulation of relationships which characterises programming environments. This articulation allows us a window onto children's learning, and the meanings which they bring to mathematical activity. Questions of interface may be important – serving what Confrey calls 'the connection between a student's current intellectual viewpoint and the underlying structures of a program' (Confrey, 1993, p. 56) – but they are secondary to the critical aspect of programming: that the learner can describe mathematical objects and relationships in a language that can be communicated, extended, and reflected upon.

Our mathematical focus leads us to concentrate on the structures, relationships and allowable actions on objects, at the expense of the figurative or physical features of the computational objects that are to be manipulated. Furthermore, any mathematical concept forms part of an intricate network of concepts, so addressing one inevitably necessitates calling upon understandings of a range of related mathematical ideas, conceptions of which will shape interactions in the computational setting. This is both a plus and a minus. On the plus side, we might hope that it is possible, by immersion within a sufficiently complex and rich medium such as programming, for such connections to be elaborated by students. On the negative side, the interrelationships and local conceptual connections make it problematic to seal off exploratory worlds into neat packages for study.

There are two further points. First, the fundamental facet of programming environments is the imperative of formalisation inherent within them. We will provide many examples of the critical role played by the linguistic formalisation of programming in aiding learners' mathematical expression and in developing their mathematical understandings. There is a degree of compact and rigorous expression involved in programming which if not isomorphic to, is at least comparable with that demanded by official mathematics. Of course, it would not do to correlate mathematics and programming merely on the basis of similarities between their notational systems. On the contrary, it is partly the *differences* between the two which give programming the opportunity to open up mathematics to diversity of learning styles and expression.

This raises explicitly the question of how meanings are created within programming environments. The notational systems of mathematics, developed over thousands of years using inert media, were designed for the establishment of mathematical truth: a purpose which suits mathematics well, but mathematics learners badly. The compactness and rigour of expression, together with the fragility of traditional formal reasoning, makes mathematical notation far removed from that which makes sense in everyday settings. It seems likely that this gap between what is expected in mathematical discourse and that which characterises everyday activity, is responsible for at least some of the difficulties experienced by children in learning mathematics. Programming too demands compact and rigorous expression but – and here the choice of programming language is critical – we shall show that meanings are established as much in relation to the output of a program, as to the elements of the program itself. Papert has remarked that mathematics is a virtual reality which happens to be possible without a computer: but virtual realities are built on feedback, and virtual reality without computer feedback is not easy!

Our second point concerns diversity of approach. If we pay careful attention to the design of Logo-based worlds, we might unlock strategies for children which are simply closed in conventional media. That is, by throwing into relief particular ways of expressing mathematical relationships, the programming environment may open strategic apertures for children – ways of expressing which are available with the computer and closed in inert representations.

4. MICROWORLDS: THE GENESIS OF THE IDEA

Our paradigm case for settings which have learning as their primary purpose is that of a *microworld*. The idea of a microworld has spread through the educational culture, to the extent that the term is in danger of losing its currency, so we begin this section with a brief overview of its etymology.[8]

The term was first used by workers in Artificial Intelligence to describe a definable world of objects corresponding to some domain of the real world in the form of a computer program. As Weir (1987) pointed out, since the real-world counterparts were typically very complex, the microworlds of those early days were 'simplified versions of reality, acting as experiments to test out theories of behaviour' (p. 12). Weir quoted Minsky and Papert (1971): '...we see solving a problem often as getting to know one's way around a microworld in which the problem exists'. So when tackling a chess problem for example '...the microworld consists of the network of situations on the chessboard that arises when one moves the pieces. Solving the chess problem consists largely of getting to know the relations between the pieces, and how the moves affect things' (Weir, 1987, p. 12).

Lawler (1987) traced the origin of the term to a report of the MIT AI Lab[9] where it was used rather casually to describe 'a possible multitude of even smaller fragmentations of problem solving domains and the cognitive schemata which might be assumed to develop from interacting with those domains'.[10] Thus microworlds were born in the AI community as a way of capturing the notion of problem solving within an arena sufficiently constrained that computers might be able to achieve a solution. It was Papert who made a small but significant change to the idea – the simple and constrained arena became part of a *knowledge domain* with epistemological significance. The seed of the idea of a microworld as a segregated world designed with genetic intent was thus planted – perhaps a serendipitous consequence of Papert's own background which combined Artificial Intelligence, psychology and mathematics.

This notion of microworld is the one which pervades *Mindstorms*. It was also taken up and analysed by Piagetian psychologists interested in mathematics. For example, Groen and Kieran (1983) saw mathematics as a key facet in general intellectual development – a view which necessarily imposed constraints upon the *kind* of mathematics which interested them; this they described as 'Piagetian mathematics'. Piagetian mathematics, they claimed, stood in contrast to much of school mathematics by virtue of the fact that the latter was essentially trivial. Groen and Kieran argued that: 'Microworlds are essentially "mini-domains of Piagetian mathematics", media for inducing the kind of spontaneous reflective abstraction that leads to the construction of new logico-mathematical structures' (*ibid*. p. 372). Additionally, microworlds were viewed as 'concrete embodiments of a domain of mathematics' (*ibid*. p. 372). From this perspective, microworlds were mathematical systems with axioms and theorems; but since this mathematical framework lies beneath the surface as far as the child's direct experience is concerned, there is a need for what Papert (1980) has termed *transitional objects* (like the turtle), standing between the concrete/manipulable and the formal/abstract. Feurzeig (1987) also focused on the objects of a microworld which he defined as: 'a clearly delimited task domain or problem space whose elements are objects and operations on objects which create new objects and operations' (p. 51). Thus a significant shift had taken place in terms of the goal of microworld development: from teaching computers to solve problems, to designing learning environments by building and combining objects 'appropriate' to a knowledge domain. As a consequence of this change in focus, the transitional object begins to take on a central role.

The clearest and most elaborated example of a mathematical microworld, is that described in Abelson and diSessa's (1980) seminal work *Turtle Geometry*, which sets out non-standard geometries as subjects of study from a computational perspective. For Abelson and diSessa, microworlds had to provide straightforward access to deep ideas by exploiting a phenomenological approach, as diSessa (1987) later described in the context of his own exploration of the turtle geometry of the cube:

> Besides a kind of density of observable phenomena... it seems that salient events... happen to be correlated with good, investigable and solvable problems.

> (ibid. p. 65)

and the playful element is present too:

> ...there are tremendous possibilities for intervention, experimenting and playing... by writing little programs for the turtle.

> (ibid. p. 65)

DiSessa contrasts this situation with that of spherical geometry, which, he argued 'is just as rich with mathematical phenomena... [but is] so subtle from a phenomenological perspective' (*ibid*. p. 65). diSessa is focusing our attention on the importance of a microworld's epistemological basis, arguing that this is what distinguishes a potentially powerful investigative environment from one less amenable to exploration. By prioritising a phenomenological approach, diSessa also pointed to a vision of mathematics learning that stretched beyond mastery of content, pointing to the importance of 'making discovery part of mathematics education... not only as a learning process... but as much or more the long term perspective on learning that we foster' (*ibid*. p. 66).

The idea of microworlds involves an intention to develop an open and investigative stance to mathematical enquiry. It retains an essentially Piagetian framework within which learning is regarded as a consequence of breakdowns – incidents where predicted outcomes are not experienced (for an elaboration of these ideas in the contexts of microworlds, see Winograd & Flores, 1988). Thus the development of a microworld involves predicting where these breakdowns might occur: at the core of a microworld there is therefore a model of a knowledge domain to be investigated by interaction with the software. Exploration is necessarily constrained but in ways designed to promote learning; knowledge is not simplified, it is recognised as complex, interrelated and evolving in action. These facets are reflected in the structures of the system, particularly in its extensibility – the extent to with the elements of the microworld can be combined, recombined and extended to form new elements. This characteristic rules out software restricted to an *a priori* set of self-contained allowable actions on a hermetically-sealed set of tools.

5. OPENING WINDOWS ON MICROWORLDS

Our focus on programming and the origins of microworlds has given us some clues as to how to design a viable microworld, as well as how to set about investigating the kinds of learning which may take place within it. As we have said, there are multiple meanings for 'the computer', and our emphasis is on the relationship of the computer with the other human and non-human elements of the educational culture. How can we set about studying this relationship?

We have several methodological criteria. We want to generate rich descriptions of learners' interactions with the computer and its associated culture. We would like to characterise these interactions, and elaborate their relationship with mathematical ways of thinking. We need to develop some measure of success in learning, to assess the influence of microworld interaction.

This last criterion is contentious. Early Logo-based research tended to adopt a learning-transfer approach which invariably ended in trivial outcomes, and the closing rather than opening of windows on the intellectual processes under study. This led many to give up altogether the search for mathematical learning outcomes, and to rely exclusively on case study and descriptive analyses. Is there any alternative? We think so, and in the following sections, we outline a methodological, structural and pedagogical framework for a viable study. In the following chapter, we present an attempt at operationalising this framework.

5.1. *The issue of transfer*

The theory behind the transfer methodology presupposes that there is a single path or at least a single right answer: and, most crucially for our purposes, that it is independent not only of the setting, the situation in which the problem solving occurs, but the intellectual as well as physical resources which are available for its solution. At the same time, there is a prerequisite that the domains chosen for testing transfer hypotheses be sufficiently simple that there exist normative knowledge that is easy to delineate. Complex situations lead to complex models of the knowledge 'encapsulated' within them, or no model at all: and in such cases, there is no epistemological basis for judging what remains invariant across two or more situations.

Let us return to the insightful critique of transfer methodology levelled by Lave (1988) which we encountered in Chapter 2. Lave cites its two key assumptions. First, that 'problems are assumed to be objective and factual' (*ibid.* p. 35). They are constructed in an alien setting, by test constructors for remote 'subjects'; problem-solvers have no choice but to solve these problems, with which they have no connection; they are problems 'for' not 'by' the solvers. Where they fail to connect (by, for example, producing 'unanalysable data'), subjects are deemed to have failed. The second assumption is essentially epistemological. It rests on the belief that transfer should take place between two 'versions' of 'the same' problem (in fact, as we saw in Chapter 2, versions can be so disparate as to stretch credulity that there is anything 'the same' about them).

This takes us to the fundamental facet of the transfer experiment: it treats as given, not only a knowledge domain and a task 'within it'; it sees tools as static and inert, their dialectical relation with the situation ignored. This is a critical failing of the traditional transfer methodology; paradoxically, it will also provide an insight into what may be preserved from experimental methodologies in general.

The inappropriateness of the transfer methodology for the study of mathematical thinking is nicely summed up by Papert's rhetorical defense of his own position:

> Psychologists sometimes react by saying, 'Oh, you mean the transfer problem'. But the author does not mean anything analogous to experiments on whether students who were taught algebra last year *automatically* learn geometry more easily than students who spent last year doing gymnastics. He is asking whether one can identify and teach (or foster the growth of) something *other* than algebra or geometry, which, once learned, will make it easy to learn algebra and geometry.

(Papert, 1972, p. 250–251: emphasis in original)

This argument is rather subtle and may not be understood any better now than it was twenty-five years ago. Simply put, Papert's argument is that programming is best seen as a means of fostering a way of thinking which might make it easier to learn 'official' mathematics. Note that this implies that programming is neither synonymous with mathematics, nor that mathematical content is necessarily learned simply by programming.

Some have interpreted Papert as implying that any study concerned with evaluating effect or outcome is irrelevant. A careful consideration proves this to be false. Papert's view is that there is nothing automatic about learning outcomes; on the contrary they involve:

i. Identification of what is to be taught and learned – something different from what is to be judged as an outcome;
ii. The construction of a setting in which this may take place;
iii. The hypothesis that this will result in making *something else* (algebra or geometry, say) more learnable.

We will be able to comment more deeply on these issues when we have provided an example of a study whose aims included 'outcome' measures, in the following chapter. For the moment, we simply state that we need not reject the quest for identifying what mathematics may be learned by programming, any more than we need reject the hypothesis that learning to read may result in learning geography or history. Instead, we need to develop methodologies which are sensitive to the complexity of the problem (which, as we have seen, the transfer methodology is not); and we need to recognise that mathematics is co-constructed by experimenters, teachers and children, with tools that enter into an active relationship with problem-solvers and their solutions.

6. OBJECTS AND STRUCTURES

We have at least two constraints on the kinds of activities we might design. First, we must try to construct situations which are sufficiently concrete[11] – as well as interesting and soluble – in the hope that students will generate tasks and sub-tasks for themselves; if their activities are restricted to simple questioning (from us) and answering (from them), we will drastically reduce the opportunities for interesting intellectual action. Second, we must remain true to the idea of reconstructability: if we design structures that allow students to (re)construct new objects and relationships out of old ones, we can increase the likelihood that we will achieve some visible and tangible representation of the state of the student's thinking and be more able to observe mathematical thinking-in-change.

Now these are research aims, not teaching objectives. And while we maintain a close relationship between the two, it would be a mistake to regard them as identical. So in the remainder of this chapter, and in the following one where we elaborate a methodology in the context of an actual study, we need some way of distinguishing between these two aims. We would like to design a microworld: but we should remember that the best example we have of a microworld is turtle geometry – surely too massive for detailed systematic observation of the kind we are suggesting. We need a fragment of a microworld, a bounded but interesting subset of activity structures, which though it may fall substantially short of the full richness of a microworld, will allow us to understand the learning that may take place, and tentatively to generalise beyond it. This should properly be referred to as a microworld fragment, but – with the caveats above – we will persist in calling such an environment a microworld.

This abuse of language is not so serious. In the case, say, of turtle geometry, not much more need be said: as Abelson and diSessa eloquently testify, the provision of FD and RT ensures that the learner is provided with just the right tools to explore differential geometry. In a sense, this offers a prototype microworld, but closer investigation of Abelson and diSessa's book indicates that it is littered with smaller sub-worlds built within the turtle geometric one – we have already encountered, for example, the turtle geometry of the cube. It might therefore provide more descriptive power to reserve the term microworld for such sub-worlds: perhaps we should have referred to turtle geometry as a 'macroworld', but it is too late for such a change.

Let us return to our first constraint: concreteness. In a microworld, the central technical actors are computational objects. The choice of such objects, and the ways in which relationships between them are represented within the medium, are critical. Each object is a conceptual building block instantiated on the screen, which students may construct and reconstruct (our second constraint). To be effective, they must evoke something worthwhile in the learner, some rationale for wanting to explore with them, play with them, learn with them. They should evoke intuitions, current understandings and personal images – even preferences and pleasures. The primary difficulty facing learners in engaging directly with static formal systems concerns the gap between interaction within such systems and their existing experience: it is simply too great. That is why computational objects are an important intermediary in microworlds, precisely because interaction with them stands a chance of connecting with existing knowledge and simultaneously points beyond it.

The computational objects which populate a microworld should maximise the chance to forge links with mathematical objects and relationships – a facet which distinguishes between a mathematical microworld and a playful, exploratory world which is mathematically uninteresting. One essential property of these objects is that they afford the learner the opportunity to move smoothly between different meanings derived from actions and language, and simultaneously to build new meanings (another version of our second constraint). As Turkle and Papert (1990) put it: 'The computer stands betwixt and between the world of formal systems and physical things; it has the ability to make the abstract concrete' (p. 346).

The idea that the computer can make the abstract concrete is not without difficulty, and we will devote much of this book to exploring what such a statement might mean. We have already suggested that language is a crucial element: it provides a way to construct relationships which renders the construction visible. In this respect, a computational world differs rather obviously from, say, a world such as Cuisinaire's rods or Dienes' blocks. Here there are objects to play with, relationships to discover, but there is no language for such reconstructions. Language must be grafted on externally, by, for example, a teacher: as Brousseau puts it succinctly, it is simply not enough to perform 'some curious manipulation with yoghurt beakers or coloured pictures and then suddenly announce that 'you have discovered the Klein-Four Group" (Brousseau, 1984, p. 116).

In short, a computational world can be *autoexpressive* – it can contain the elements of a language to talk about itself.[12] The definition of what such a language might be like is broad, and broadening all the time: but we are not prepared – at least for the time being – to let go of it altogether: at root, it is the language, the program, which allows the most obvious link between computational and mathematical discourses.

The pedagogical corollary is to attempt to construct learning environments in which mathematical formalism becomes as familiar as everyday situations. An anecdote of diSessa's illustrates what we mean. He recalls designing a microworld for optics based on ray tracing – a computerised version of a formalism which describes optical systems by tracing the path of light through it (see diSessa, 1988a). As it stood, the microworld was 'useless'; it did not have 'any of the experiential feel of optical phenomena in the real world'. Yet the introduction of a single new feature changed the whole feel of the environment: 'in addition to placing optical objects down, one can place *things to see* and can ask the system to show you what is seen from any vantage point. All of a sudden abstract questions become experiential and immediate' (ibid. p. 64). The idea, argues diSessa: 'is not to juxtapose experiential and formal points of view as above, but to fuse them' (*ibid.* p. 64), to develop 'semi-formalisms', which are general enough to express ideas, but which retain control and familiarity for the learner.

DiSessa's semi-formalisms stand between formal and informal, systematic and intuitive: they help us to understand what it means to 'make the abstract concrete'. This is precisely what mathematical formalism does for the mathematician: the process of algebraic formulation is the means by which mathematical objects (concrete for the mathematician) are manipulated. An autoexpressive environment aims at conferring this expressive power more broadly, by making the algebra of relationships between things (and relationships) semi-formal, concrete, meaningful.[13]

6.1. *The question of pedagogy*

Let us recall Papert's hypothesis: he asks whether we can teach something *other* than mathematics, which will make it 'easy' to learn mathematics. Two issues come immediately to mind: what kind of pedagogic intervention is appropriate; how explicit should and could it be? As far as explicitness is concerned, one view might be that by immersion in a culture of sufficiently rich, interactive and motivating experiences, children will acquire and come to appreciate the salient features of the microworld – whatever these might be. In this perspective, the teacher's role during microworld interaction would be minimal. The French didacticians describe such situations as *adidactical*: students take on a problem for themselves without explicit teacher intervention, but the problems are carefully designed so that they are likely to call upon mathematical tools in their solution. The idea of adidactical situations – such as microworlds – is that students exploit and reconstruct mathematical ideas *by choice* and not as a result of teacher insistence. The computer is especially helpful in this regard:

> ...the fact that students interact directly with the computer without direct input from the teacher is not sufficient to guarantee the success. It is only by means of a negotiation which aims at the devolution to the student of the control of his or her interaction, that the quasi-isolation of the student from his or her teacher is obtained. The didactical contract ruling the situation will allow it to be close enough to a situation 'free' of the teacher's influence.

> (Balacheff, 1993, p. 156)

The idea of adidactical situations converges with the view of pedagogy found in much of the LogoMathematical literature. Unfortunately, the latter has frequently been interpreted as advocating no intervention at all. Apart from its simplistic viewpoint, this reading fails to appreciate the elements of pedagogy present through other agencies: how the teacher/researcher sets up the situation, the materials which are present, the activity structures which are devised.

The zero-intervention strategy is not viable. As Harel (1991) puts it, knowledge must be built both 'internally and externally', by one's own construction and by a knowledgeable adult: the computer cannot substitute for the latter. We require more precision as to what teachers might say and do during microworld interactions, and how they are contingent upon students' actions. The situation is made more complex by our insistence that mathematics is not a unitary practice restricted to lists of contents: our pedagogy will need to take account of introducing students to the rules of the mathematical game.

We know that people can be taught little chunks of mathematics and can repeat them if tested a short time later. We also know that they are unlikely to be able to mobilise this knowledge, nor use it effectively in problem-solving situations. Following the French researchers, we can conceptualise this phenomenon as the *didactical paradox*: didactical situations 'force' teachers to tell the students what to do; yet this same process empties the learning situation of any cognitive content – the students will not and cannot learn as they obtain solutions by interpreting the didactic contract rather than engaging with mathematics (see Brousseau, 1986).

In summary, we see every reason to be explicit that in MicroworldMathematics there is a new discourse, with new objects to attend to and things to do. At the same time, we need to allow enough freedom so that students can own problems, and explore them for themselves. We want to borrow the idea behind the adidactic situation, but we cannot rule out intervention. In the next chapter, we will illustrate how the computer can make this possible; the central idea is that MicroworldMathematics affords a depth of engagement and strength of expressive power which can provide a neutral buffer between teacher and taught.

Our emphasis on exploration and play, faces us with a version of the didactical paradox, which we have called the *play paradox*. When we are playing, immersed in a game, we are attending to the experience of play itself: where to hit the ball, how to move the chess piece, when to strike the drum. We may have, at a more or less conscious level, a rich set of rules about the discourse of the playful situation in which we are immersed. We may be learning more about these rules, and gaining experience of the way they are operationalised. But there can be no guarantee that we will learn just what someone else plans we will learn. To maximise this possibility, it might be that we need to be told: but then we are not playing. If the teacher brings a new idea to the student's attention from *outside* the activity, then he or she is no longer playing: yet, if it is not imported, he or she might never encounter the idea. Play confers meaning to an activity but blurs the specificity of the intended meaning.

Thus the balance between exploration and guidance is always problematic.[14] We want learners to have sufficient time and space to nudge up against the learning goals of microworld environment, to play with its objects and relationships, and to take on board the tools as a personal medium of expression. Yet we also want them to reflect on the structures from a mathematical point of view, and to focus their attention on the pieces of a mathematical puzzle which we may have carefully planted within the activities.

This tension is not completely resolvable. We might be able to engineer situations in which a mathematical way of thinking is encouraged. But mathematics *per se* is not discovered by accident. Any new research setting we devise should provide opportunities to study this tension more closely.

Let us sum up our aims for this setting, whose elaboration takes up much of the next chapter. Our twofold aim is to mobilise the computer as a window on to children's thinking and to offer a means by which children can express themselves mathematically. This duality spills over into our design criteria for microworlds. Our interventions on a pedagogical level reflect this duality: we see a dialectical relationship between intervention designed to explain, elaborate, and connect, and intervention designed to probe, understand and investigate. Intervention which 'tells' or attempts to 'transmit' is not only unlikely to be successful, but runs counter to a primary design criterion. We are unlikely to learn much about children's thinking if we design intervention strategies which render us deaf to what they say.

From a research perspective, the careful description of intervention is often neglected. So an important component in what follows consists of a description of pedagogical structures which map a mathematical perspective onto the activities within the setting, shaping students' work while at the same time being contingent upon it. There will need to be both off- and on-computer activities, and a rather careful consideration of the sequences of tasks which intersperse exploratory or playful interactions with more focussed sequences of activities.

At the same time, we will need to recognise the importance of considering the students' views of their activities from which meanings are derived: these are often unanticipated and unexpected. Students' preconceptions and intentions will fundamentally affect how they think of their activities, the strategies they adopt and their interpretations of the results of their actions. We will need to understand what learners think about the ideas on offer, and listen to the ways in which these are re-presented during activities. Here is a final, methodological, tension: between structuring activities which allow learners to generate meanings for themselves, and the revelation (to us, and to each other) of the meanings, strategies and intuitive mathematical frameworks they have created.

NOTES

1. In Chapter 2, we explained that the dynamic geometry system we refer to in this book, is Cabri Geometry (Baulac, Bellemain, & Laborde, 1994). This, like other systems, introduces a dynamic element into geometry – fundamental Euclidean elements of a figure (e.g., points and lines) can be grasped with the mouse and dragged around the screen. The movement keeps intact any geometrical relationships which form part of the figure's construction together with all the properties dependent on them. Thus these relationships can be *seen*: they are invariants in the dragging process (see Laborde, 1993a; Laborde & Laborde, 1995).

2. ...another dialect of Lisp.

3. So too is a photocopier. For a fascinating account of the design issues involved see Suchman (1987).

4. This example nicely illustrates the danger of such dichotomies. Mathews (1989) reports that on a flight from London to Bangkok, the pilot is in control of the plane for something like eight minutes: the rest of the time, the computer is in control. Airplanes often fly with their crew asleep, and it is not uncommon for planes to overshoot their destination in this state. Göranzon's forest rangers, whom we encountered in Chapter 2, illustrate much the same phenomenon.

5. This has been tried, with some success, in experimental schools in Bulgaria.

6. *LogoMathematical* is a term which, we hope, will acquire a sufficiently rich set of meanings by the time we are finished.

7. The idea of a *microworld* is the main focus of the following section.

8. This section is based on Hoyles (1993). For further discussion, see Edwards (1995).
9. 1972 Progress Report by Minsky and Papert.
10. Lawler himself described microworlds as socio-cognitive schemata with specific characteristics of activity and interaction. He adopted the term, *microview* in Lawler (1985), for his more psychological formulation: Papert was developing a rather different meaning. Lawler argued that a distinction needed to be made between what is in the world and what is in the mind as well as between the experience of the expert and that of the novice. He coined the phrase 'miniworld' to refer to 'the object-focused embodiment of some designed environment, as described by an expert in the domain; and defined microworlds as 'partial exploitations of the complete generality of what might be possible in the miniworld'. Microviews were defined as the cognitive structures developed from novices' interactions in microworlds. Thus, in Lawler's conception, understanding the idea of the miniworld follows after interaction with more than one microworld and the coordination of the associated microviews.
11. Recall that this means 'well connected to what the learner already knows'.
12. This also points to a difference *between* computational worlds. For example, Jim Kaput (1992) describes 'cybernetic manipulatives' which can be combined, constructed, and reconstructed; but he envisions no explicit role for a language in which to express these constructions.
13. We are still using these terms interchangeably, and without sufficient clarity. One aim of Chapter 5 is to clarify their meanings and relationships.
14. For a discussion of how this balance might be achieved in a more directive educational system, see Kynigos (1995).

CHAPTER 4

RATIOWORLD

We now describe a microworld for working with ratio and proportion,[1] based on the design principles we set out in the previous chapter. We summarise our strategies for investigating what children learned and our reflections on the study in the light of the theoretical framework we have outlined so far.

1. WHAT DO WE – AND LEARNERS – KNOW?

Ideas of ratio and proportion form a rich domain of mathematical structure, and one which is well researched in the psychological niches of the mathematics education literature. Researchers have provided a comprehensive picture of a range of student responses to ratio questions which can be expected in conventional, non-computational environments and studied within traditional methodologies. For example, Inhelder and Piaget (1958)[2] classified four 'stages' of development of the concept of proportionality, beginning with an inability to focus on more than one variable at a time, and ascending to a final (formal operational) stage in which two variables are coordinated. In accordance with Piagetian theory, these stages are deemed independent of problem setting.

Research since then has indicated that the picture is not so straightforward, and that children's understanding of proportional relationships depends on a variety of factors.[3] Some of these kinds of findings can be summarised by the CSMS[4] study:

   (i) the methods used by students to solve problems vary with the problem presented;

  (ii) most students see ratio as an additive operation and essentially replace multiplication by repeated addition; and

 (iii) in enlarging figures 'there is the danger of being so engrossed in the method to be used that the student ignores the fact that the resulting enlargement should be the same shape as the original'.

(quoted in Clarkson, 1989, p. 191)

These and other studies provide valuable baseline data on children's ideas about ratio and proportion, and point to the limitations of a strictly Piagetian approach. However, following our theoretical discussions in Chapter 2, we might note that there is relatively little recognition accorded to the potential influences of task, setting and tools on the construction and mobilisation of ideas of ratio and proportion.

At the same time, we will need to take account of teaching strategies as well as students' performance. The addition strategy described in (ii) above, has often been interpreted as a misconception and accordingly there have been teaching experiments aimed at challenging it. Hart (1984), for example, set out to 'eradicate the use of the addition strategy' using teaching material that relied on the calculator. Her research reported success, provided the teacher was 'sensitive to the methods employed by the students and the reasons for their use' (*ibid.* p. 81). In a later study by Johnson (1989), which also included the investigation of ratio and proportion, the researchers aimed:

(i) to investigate the development in students of formal mathematical methods and conceptual models with particular attention given to classroom experiences;
(ii) to ascertain the connections between the information assimilated by the student and that given by the teacher;
(iii) to identify the nature, use and development of 'student methods'.

(ibid p. 4)

One of the major findings of this work concerned classroom teaching; that teachers consistently failed 'to take account of what ideas and strategies the students had *before* the start of the teaching programme' (emphasis in original, Johnson, 1989, p. 217). This leads us to wonder if the findings concerning students' failure to adopt a mathematical perspective on ratio and proportion, the 'gaps' observed between pre-formalisation and formalisation, may be due to the kind of teacher/student interaction on offer, coupled with something more profound in the culture – a lack of connection between the practice of mathematics and everyday practice. We know that proportions are a major stumbling block for many children; investigating why, will lead us to consider a range of dimensions of the problem which take account of the paucity of children's experience with multiplicative structures and the different types of situation which require multiplicative relationships. It will turn out that this will be the first of many intersections of the cultural and psychological domains which will afford insight on mathematical meaning-making in general.

This brings us to a consideration of the knowledge domain[5] itself. Vergnaud 1983), while recognising the problem-oriented nature of responses to tasks, has identified the *conceptual field* of multiplication within which he cites three subtypes of multiplicative structures. One of these is the 'isomorphism of measures',[6] which consists of a simple direct proportion between two measure spaces. A good example is the relationship between a number of items (n) and their total price (p). In this case, the relationship is fixed, so for a different number of items f(n) say, there is a corresponding new price f(p) where f is a multiplicative function. There are only two variables (n and p) and there is a one-one mapping between them. Carraher, Schliemann & Carraher (1988) noted in their review of the literature on proportional thinking, that 'the isomorphism of measures model seems to represent well the way (some) people think about multiplication and proportion problems in (some) everyday activities' (*ibid.* p. 26).

Within this subtype, the solution to a general multiplicative problem can be obtained by the use of a scalar operator *within* a measure space M_1 or M_2, (σ in Figure 4.1), or by the use of a function operator (ϕ) *between* measure spaces. That is, a problem-solver can think about the relationship vertically ($n_1 : n_3$ so $n_2 : ?$) or horizontally ($n_1 : n_2$ so $n_3 : ?$). This situation is clarified in Figure 4.1:

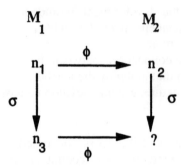

Figure 4.1. Scalar (σ) and Function (ϕ) Operators within and between measure spaces M_1 and M_2.

There is a complexity here which is interesting in terms of the relationship between everyday and mathematical ways of thinking. One of the most revealing insights on the problem is derived from Nunes et al. (1993), who argue that this isomorphism of measures model (in whichever variant) 'does not immediately apply to other types of [proportional] situations' (*ibid.* p. 81). It is however, the one which most people adopt initially for two reasons. The first is that it is clearly the simplest as it minimises the number of variables involved. The second, deeper, reason is that 'the understanding of the situation is kept in view when the mathematical relations are analysed' (*ibid.* p. 81). That is, each element of the chosen calculation maps directly on to a corresponding element of the problem, and the mental schema which is developed 'includes the situational meaning and their mathematical relations in the same representation' (*ibid.* p. 83). This stands in marked contrast to other, more complex models.

The difficulty is that the isomorphism of measures model is limited in its applicability. For example, the model works well when the value of the variable in M_1 is increased, and allows, say, a repeated addition strategy to be invoked. However, it does not work when the value is decreased: then there needs to be an explicit recognition that a multiplicative relation is involved, and that its inverse – division – is required. As Nunes et al. put it, this has interesting repercussions:

> ...if situational information and mathematical relations are inextricably interconnected in the model, it becomes impossible to envisage a general mathematical model that can be applied to other situations.

> (*ibid.* p. 83)

From a psychological point of view, this is interesting enough, as Nunes and her colleagues amply demonstrate. From a pedagogical point of view, it can generate a number of responses, the most likely being to encourage the learner to break conclusively with the situational information of the model and concentrate exclusively on the mathematical relations involved: in other words, to sever the connection with the 'concrete' and contextualised and ascend to the 'abstract' and decontextualised.

It will be clear from all that has gone before that, while we cannot totally reject the intention behind this latter approach, it will not suffice. Our aim now is to construct an alternative approach by studying what happens when we insert an appropriate microworld into the teaching/learning setting. In order to base our alternative on more than mere speculation, we provide details of a study we have undertaken of children's proportional thinking within a computational microworld.

2. BUILDING AN ALTERNATIVE METHODOLOGY

We start from the premise that each individual has a complex set of understandings of the idea of proportionality, and that any learning situation should aim to set in motion the reconnection of these understandings with each other, and with new understandings derived from interaction in the situation. Studying thinking-in-change, we can begin to analyse what varies and remains invariant across the situations we design. More fundamentally, by analysing shifting rather than inert states of understanding (or more exactly, performance), we have a greater chance of gaining insight into what is known, how that knowledge is mobilised, and the meanings which an individual constructs as new knowledge comes into being.

In this study we designed a sequence of activities which varied according to setting (individual, small group and class) and the tools available (paper-and-pencil and computer). The sequence, which we call *RatioWorld,* comprised: a series of individual written tasks, six one hour sessions around microworld activity (computer and non-computer based), followed immediately, and again some time later, by the same series of written tasks. Our objective throughout the entire sequence was to introduce students to a *proportional way of thinking*. Below we give an outline of the different activities before we turn to the details of the study.

2.1. *The written tasks*

The tasks were divided into two parts. The first part consisted of a *recognition* question involving the identification of a set of given rectangles which represented plans of swimming pool of various dimensions: we wanted to know how students classified 'equivalence' of rectangular shapes and what criteria they used.

Jo and Pat are painting the garden shed. They want it to be grey. The shop only has small tins of white paint and small tins of black paint. Jo mixes ten tins of white paint with fifteen tins of black paint. Pat has two tins of white paint. Here is a table showing these amounts:

	Amount of white paint	Amount of black paint
Jo	10	15
Pat	2	?

How many litres of black paint must Pat use to get the same shade of grey? Pat needs .... tins.

Figure 4.4. A written task. Paint: scalar integer.

The second part involved the students in the construction of an unknown value given two measure spaces and three initial values. Each situation *could* be solved by seeing it as a problem of proportionality. We set our tasks in two distinct contexts: the first concerned mixing paint *(paint context;* see Figure 4.4), and the second enlarging photographs of rugs *(rugs context;* see Figure 4.5). We were interested in whether these variations in situational information would have any influence at all on students' constructive and verification strategies – and if so in what way. We also wanted to investigate whether there were differential shifts in response depending on whether the questions were numerical *(paint)* or graphical *(rugs)*. At the same time, we were concerned to probe the influence of the type of *scale factor*: we know, for example, that integer scale factors are deemed to be easier in written tests than rational ones. We therefore included tasks which reflected these and related distinctions, (e.g., the possibility of a non-integer answer appearing as a solution given an integer scale factor in a divisive rather than multiplicative role).

Finally, we wanted to probe further the evidence that despite formal logico-mathematical equivalence, some situations afford a 'direction-in-action' which appears to preserve the semantic meaning. So we set out to explore the interaction between context (rug/paint) and a preference for function or scalar operators (within or between measures).[7] For example, did it make more sense, in the question shown in Figure 4.4, to think of the relationship

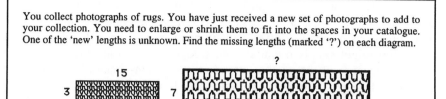

Figure 4.5. A written task. Rugs: function integer.

between the numbers of tins of white paint (*scalar*), or between the numbers of tins of white and black paint, mixed by Jo or Pat (*function*)?.

In summary, we had *three* variants of our tasks: *context* (rug/paint), *scale factor* (integer/fraction[8]) and *operator* (function/scalar). We were therefore able to designate *a priori* what kind of task each was (e.g., 'function integer'), although there was not, of course, any necessary relation between this designation and the chosen strategy of any student. Table 4.1 illustrates the distribution of the three variants over the ten tasks.

Table 4.1. Classification of questions: combinations of the three variables context/scale factor/operator.

Context	Scale Factor
Paint	Function Integer
	Scalar Integer
	Fraction > 1
	Fraction < 1
Rug	Function Integer
	Scalar Integer
	Integer division
	(decimal solution)
	Fraction > 1
	Fraction < 1

2.2. *The computational objects*

Following the written tasks we designed several activities, which we will describe later. Here we focus on the technical element of *RatioWorld*, its computer-based components. Our aim was to try make concrete the mathematical idea of 'proportionality';[9] it was to be a place in which it made sense to express multiplicative relations generally and formally. In short, we were attempting to open a window for students onto proportional thinking.

Seeds of our design principles can be found in several earlier research studies. First, the potential of the turtle to synthesise symbolic and graphical representations seemed to offer a medium in which the difficulty referred to by Hart might simply not arise: i.e., students 'forgetting' the shape to be enlarged because of the need to attend to other, non-visual elements of the problem. More generally, our experience had shown that working with a turtle tended to throw up ideas of ratio and proportion rather naturally – students became part of a multiplicative world. For example, once procedures for drawing figures have been built, students often pose for themselves the issue of enlarging or shrinking them. Growing and shrinking do not, of course, necessarily involve proportionality. Nonetheless, the kinds of Logo activities in which students engaged in a relatively unconstrained way, could frequently be seen to mirror the isomorphism of measures model reasonably well. Hoyles (1989) for example, showed how students used the Logo idea of *input* as a scale factor (a scalar operator) to change the size of a drawing in proportion (*ibid.* p. 198), and built general procedures in ways that reflected the internal relationships between figures (function operators) (*ibid.* p. 190).

The second seed is to be found in the N-task study described in Chapter 2. It provided us with some experimental evidence that working with Logo could make a qualitative difference to what students did with proportional ideas. In particular, we knew that learners sometimes adopted strategies and exhibited skills in the computational context that were different from those in paper-and-pencil contexts, and that the formalisation within a specific Logo solution had the potential to capture more general proportional relations in a problem.

As well as the turtle graphics subset of Logo, we also wanted to incorporate Logo arithmetic as part of the microworld activity, as a means to do proportional calculations. This, we hoped, would help students avoid missing the proportional wood for the calculational trees and counteract the tendency for 'sharp increases in difficulty when the multiplier changes from an integer to a decimal number....' (Bell, Greer, Grimison, & Mangan, 1989, p. 447). While we had no expectation that the provision of Logo to perform troublesome calculations would automatically cancel out all these difficulties, we did hope that it might at least provide support for them to the extent that we could study more interesting aspects of learning.

Two computational objects formed the basis of the technical component of the microworld. We conceived of them as transitional objects, conceptual building blocks which students could construct and reconstruct, and in so doing, engage in thinking and talking about proportions. The first was HOUSE, a fixed procedure drawing a closed shape (see Figure 4.6). The lengths of the sides of HOUSE were not simply related – there was no obvious common unit or multiplying factor connecting them.

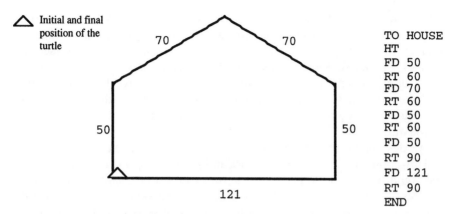

```
TO  HOUSE
HT
FD 50
RT 60
FD 70
RT 60
FD 50
RT 60
FD 50
RT 90
FD 121
RT 90
END
```

Figure 4.6. HOUSE, a fixed closed shape.

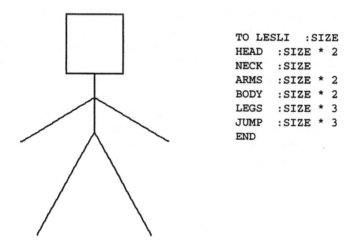

```
TO  LESLI   :SIZE
HEAD    :SIZE * 2
NECK    :SIZE
ARMS    :SIZE * 2
BODY    :SIZE * 2
LEGS    :SIZE * 3
JUMP    :SIZE * 3
END
```

Figure 4.7. Pinfigure LESLI: a shape built according to proportional relationships.

The second was a general procedure (i.e., a procedure with an input) LESLI, which drew a pinfigure (see Figure 4.7). LESLI was made up of variable subprocedures, pre-written according to proportional relationships.

Our intention was that students would play with HOUSE and LESLI and generate for themselves new problems and solutions based on the figures, as well as investigate the effects of modifying them in ways which we suggested – to keep everything *similar* in a mathematical sense. The analysis which follows is based entirely on HOUSE: details of the LESLI activities are subsumed into the findings.

As well as these two central objects, we provided pre-written tools JUMP and STEP which, respectively, moved the turtle (without drawing) up and across the screen (if used with positive inputs). We did this in order to try to bypass issues of turtle orientation and turtle turn which are sometimes sources of confusion for students with limited Logo experience (see, for example, Hillel & Kieran, 1987; Hoyles & Sutherland, 1989, Chapter 8). We wanted students to focus attention on what mattered to us, rather than allowing potentially misleading avenues to be pursued – although, of course, students were free to manipulate the turtle using RT and LT if they wished (some did).

We designed a mixture of exploratory and structured activities, both on- and off-computer, including pair-work and small-group and class discussions. We could not avoid the pedagogic tensions raised in Chapter 3, in particular, the thorny question of when and how to intervene over and above the intervention implicit in the activity structures themselves. We recognised that in a classroom situation with many computers, there was a need for clear management and organisation. Nonetheless, we tried to tread carefully the boundary between structuring activities which would induct students into ways of thinking we wanted to encourage, and at the same time giving them the freedom to explore ideas and pose problems for themselves. In the description that follows, it is rather easy to illustrate the former, and rather difficult to show the latter. It is however worth reiterating what we didn't do. We did not attempt to encourage 'transfer' between written and computer tasks (or *vice versa)*; we did not teach rules for working out proportion questions in any explicit or formal sense (such as the 'rule of three'); and we did not teach students how to answer the questions in our written tasks either by pointing out that there *was* an underlying proportional structure to these questions or by teaching a technique.

3. THE DATA SET

In this section, we introduce the students with whom we worked and outline the methods of data collection we adopted.[10]

3.1. *The students*

We worked in one school with a whole class of 28, 13-year-old students, recognised by the school as being in the 'middle range of mathematical attainment'. The students worked on the activities for one-and-a-half hours per week over a period of 6 weeks. The class's mathematics teacher was experienced both in mathematics teaching and in using the computer in her class; she used calculators as part of her normal classroom activities. The students had some experience with Logo,[11] and were relatively familiar with simple Logo-turtle syntax. None of the students had been formally taught ratio and proportion in their mathematics classes.

We assumed responsibility for teaching the class, for setting and marking homework assignments and giving feedback to the class teacher. The class teacher was present at each session and gave us feedback so that we could gain a feel for the pedagogical constraints of teaching a whole class in school, as well as put ourselves in a position to respond to questions from the students. She assisted us in our observations and helped us to interpret individual students' activities. We also gathered some baseline data from a number of classes of students in the school who did not participate in the activities (a 'baseline group'[12]) in order to provide a backdrop against which the microworld activities of our class could be assessed.

3.2. Data collection

We wanted to attempt a systematic evaluation of thinking-in-change over our sequence of activities. In order to achieve this, we adopted an eclectic and broad range of methodologies. First, to begin the process of identifying trajectories of change in the states of individual students' understandings, we collected students' responses to our written tasks before, immediately after, and some time after the microworld activities; we will refer to these as pre, post, and delayed-post tasks. In addition, we gave the same set of tasks at the same three times to the baseline group. We planned to analyse this data set statistically in order to identify trends and raise questions, so that we could subsequently seek answers from a range of qualitative data collected by close observation and case study of a group of eight students.

These case-study students were chosen so as to obtain a spread of mathematics 'attainment' (as judged by the mathematics teacher), an equitable mix of girls and boys, and a range of responses to the pre-task questions. We audio-recorded interviews with them subsequent to the pre- and post-tasks, in order to try to understand how students had arrived at their responses to the written tasks: how, for example, had they interpreted the meaning of the question? Had they used any form of measuring instrument (e.g., a ruler) in the recognition question? Which operation had they chosen in the *paint* and *rug* questions, and why? Had the students used the situational information of the questions, and if so how? Did they check the feasibility of their responses against the situational information of each task?

The interviews were further aimed at teasing out those elements of the mathematical structure of the questions (if any) that the student had attended to, and the extent to which they adopted a proportional way of thinking about them. A crucial facet was their preparedness to see the need for any strategy which was in some way consistent within a context, or even across contexts. In cases where we knew students had adopted inconsistent strategies in a given context, we offered them back these strategies (producing two different answers), and asked if both answers could be correct – if not why not. A useful strategy we employed was to ask:

If your life depended on it, which, if any, response are you certain is right and why?

Finally we needed detailed, rich descriptions of the students' responses during the microworld activities, and to this end we observed the progress of our case study students closely: we noted what they said and did, collected their homework assignments given after each session, the printouts of their Logo procedures and the dribble files of all their computer interactions. Where possible, we monitored their thinking and interpretations of the activities through clinical probes during the microworld activities (for example, asking why they had chosen a particular strategy).

4. THE MICROWORLD ACTIVITIES AND STUDENT RESPONSES

The six sessions of the microworld consisted of a mixture of computer-based and off-computer activities: the latter were divided between class activities, class discussions and small group work. In all, the activities were structured around ten tasks. We present a subset of the activities here – a full description of them can be found in Hoyles, Noss and Sutherland (1991b) – pausing to provide a little more detail where necessary, and sketching some relevant episodes of the children's responses.

4.1. *Meanings of similarity*

We started by asking groups of students to sort pictures into classes on the basis of similarities they perceived, and to work out a justification for their sorting which they reported back to the whole class. Our hope was that in doing so, students would reveal and make explicit in natural language their intuitions of similarity, and provide us with an opportunity to highlight the differences between the use of 'similar' in everyday parlance and its more specialised mathematical sense. We then moved on to a more focused activity – the I-task – involving comparisons of different size and shapes of the letter I, in which a range of possible meanings for the idea of 'in proportion' (collected from our pilot work) were offered to the class for discussion. This gave us and the students the chance to look at the validity of the approaches, and what was special about each (including, of course, the 'mathematical' meanings).

We were not surprised to find that students were very likely to adopt a classification of the figures we presented based on one or two attributes, rather than on the considerably less tangible relationships *between* attributes. It was also clear that the expression *in proportion* had currency in natural language, where it assumed a variety of meanings including 'roughly equivalent', or 'about the same shape'.

In the I-task, we tried to push students a little further in the direction of appreciating the difference between everyday and mathematical discourse. A

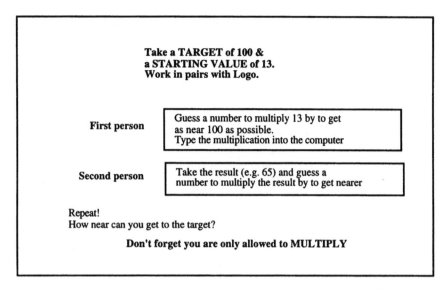

Take a TARGET of 100 &
a STARTING VALUE of 13.
Work in pairs with Logo.

First person | Guess a number to multiply 13 by to get as near 100 as possible. Type the multiplication into the computer

Second person | Take the result (e.g. 65) and guess a number to multiply the result by to get nearer

Repeat!
How near can you get to the target?

Don't forget you are only allowed to MULTIPLY

Figure 4.8. The Target Game.

typical response was that of Toby, one of our case study students, who began very uncertainly, and was unable in the classification task to choose between different strategies he had employed to sort similar objects – any strategy was acceptable provided 'it looked OK'. When first comparing different I-shapes, Toby thought that two I-shapes were in proportion *as they were the same letter*; later his decision was based on whether the same amount had been added to each part of the smaller one to obtain the larger.

During the two initial activities, we saw Toby beginning to appreciate that mathematical relationships were different from everyday judgements. Being the same letter (I) is of incomparably more importance in everyday discourse than the internal or external relationships between the parts of the letter. Mathematically (at least in the scenario we set) the reverse is true, and it is this, rather than learning any specific technique, that we think Toby had begun to appreciate.

4.2. *Broadening the notion of multiplication*

In our next session, we tried to gain some insight into the students' sense of multiplication by non-integers, asking them in pairs to play 'The Target game' which used Logo's arithmetical operations.[13] This activity is reproduced in Figure 4.8.

On one level, we incorporated this activity in recognition of the difficulty we anticipated the students would encounter. The belief that 'multiplication makes bigger' is widespread, arising from a multitude of common experiences: we already knew that many students shared this belief. So from this point of view, we wanted to offer situations in which multiplication, once employed, could *not* make bigger. More generally, we wanted to set up a situation in which operating multiplicatively on numbers was imbued with richer-than-everyday meanings, and in which the results of such operations were both immediately evident and could be retained for collaborative reflection.

As it turned out, the effect of multiplication by numbers less than one (and of multiplication by a negative number – apparently a surprise for many students) successfully challenged and expanded several students' notion of number. For example, although Jenny had answered correctly on the I-task, she was initially confused between negative and decimal inputs to primitives and procedures when working with the computer: this opened an important window on her understandings. The I-task required her to interpret *our* meanings, but when she had to *construct* for herself using Logo, different, previously hidden facets of her appreciation of proportionality became evident. Perhaps in the teacher-guided discussion she had previously simply 'guessed our agenda'?

In her response to the Target Game, Jenny appeared to be developing new ways of seeing number and number operations. Jenny, and her partner Sue, were trying to reach a target of 100, with a starting value of 13. Sue started by multiplying 13 * 8 giving 104. Jenny's first response was to multiply (she knew the rules of the game!) but by – 4. Much laughter. Then the two girls decided to start again. Jenny again tried 13 * 8, and Sue, using 104 as a new starting point, tried 104 * 0.25. The sequence continued with 26 * 3.75 (Jenny) then 97.5 * 1.2 (Sue). Jenny was then faced with 117. She tried 0.2 – the first time she had chosen to use multiplication by a decimal between 0 and 1 to reduce a number. A fragment of the pair's attempts is given below:

```
23.4 * 4.25
99.45 * 1.01
100.4445 * 0.99
99.440055 * 1.0001
99.4499990055 * 1.001
99.549449004505 * 1.002
99.748547902514 * 1.002
99.948044998319 * 1.0000001
99.948054993123 * 1.000001
99.948154941178 * 1.0004
99.988134203154 * 1.00045
100.03312886354 * 1.00025
100.5813714575 * 0.00005
100.05813714575 * 0.9999995
100.05808711668 * 0.999995
```

```
100.05758682624  *  0.99995
100.05258394689  *  0.9995
100.00255765491  *  0.9995
```

It appears that the pair were developing a growing confidence and awareness of the effects of decimal multiplication. Our only interventions were at the beginning of the game to clarify its rules – division was not allowed and the students had to take turns in their attempts to reach the target. It is worth noting that after these activities, Jenny was 100% correct in her homework questions, an example of which was 'What do you multiply 80 by to give 30?'.

We have to be cautious about attributing one activity to the growth of one particular understanding. Yet the computer offered something new to Jenny: an aperture through which an alternative strategy – an iterative solution – made sense, and whose steps could become objects of reflection as they were stored on the computer screen. Indeed, this aperture was effectively all we offered: by exploiting it we merely manipulated the situation to increase the likelihood of Jenny developing the multiplicative insights we desired.

4.3. *The imperative of formalisation*

We gave the students the procedures for HOUSE, STEP and JUMP, and simply asked them to build any configurations they wanted to explore: bigger and smaller HOUSEs, rows of HOUSEs, abstract patterns of HOUSEs and so on.

Next, we asked them to build bigger and smaller HOUSEs *in proportion*, encouraging them to make their strategies explicit through writing programs. This enabled them, for example, to build on their appreciation of 'doubling' by expressing it as '* 2'; and we nudged them further to generalise to other multiplying factors. This raised an apparent paradox. We built this activity around the idea of *doubling*: yet we knew that many students see doubling as a special case of addition, rather than as a multiplicative operation. In fact, in our earlier versions of the activity, we explicitly ruled-out doubling as a 'legal' strategy, and based it around a task which effectively outlawed doubling altogether. After our trialling, we realised that this approach failed to take account or exploit new tools available in our microworld. We had unthinkingly simply transferred a paper-and-pencil finding to this new context: while starting from doubling in an inert medium might encourage the student in his or her belief that multiplication was not necessary, the dynamic medium with which we were working could be used to point out quite the opposite. We adopted a simple but effective approach: to start from a strategy which students adopted intuitively, doubling, and to use the imperative of formalisation inherent in the medium to encourage students to generalise from doubling (formalised as * 2), to trebling (* 3) and, finally, towards operating with a general multiplier (* :x).

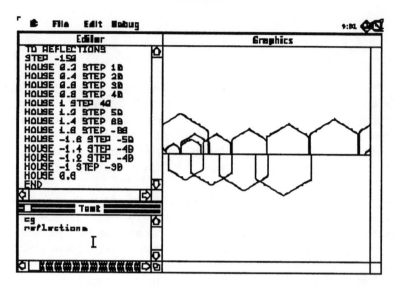

Figure 4.10. Russell's reflections.

This worked extremely well for almost all our students, but it is worth pursuing in slightly more detail the strengths and limitations of the microworld approach. Consider Russell, who was happy simply to take the inputs to FD of the HOUSE and multiply by any integer – or even by a decimal in simple cases. He built HOUSE :P (with help) where :P was the scale factor, and used this to create a 'reflection' pattern with a range of inputs for HOUSE (see Figure 4.10).

However, analysis of Russell's paper-and-pencil responses *after* this computer activity revealed inconsistencies: he variously used multiplicative and additive strategies in his homework problems. One explanation might be that on the computer, Russell had used decimal inputs between 0 and 1 to his variable HOUSE procedure to obtain his required visual outcome (smaller HOUSEs), but had *not* discriminated the mathematical 'necessity' of the multiplicative operation. This discrimination was *our* agenda, not Russell's: thanks to the ad-hoc nature of our interventions – inevitable in a class situation – we had simply failed to make an appropriate intervention during Russell's playful activity at the computer.

A second interpretation might be that Russell had difficulty in operationalising his multiplicative strategy in a non-computational setting: the homeworks could only be done without a computer, so that the visual outcomes of running the Logo procedure were unavailable to assist him in problem solution. This might, of course, not be simply a matter of feedback: it might be that there is something deeper at work, and that the means of expressing the relationship was closely bound into the setting in which it was articulated.

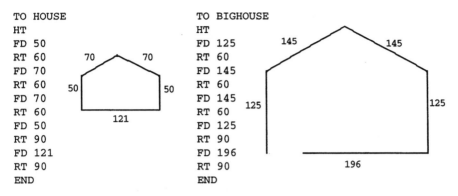

```
TO HOUSE                TO BIGHOUSE
HT                      HT
FD 50                   FD 125
RT 60                   RT 60
FD 70                   FD 145
RT 60                   RT 60
FD 70                   FD 145
RT 60                   RT 60
FD 50                   FD 125
RT 90                   RT 90
FD 121                  FD 196
RT 90                   RT 90
END                     END
```

Figure 4.9. Using addition to enlarge HOUSE. The strategy is evident in the code (75 has been added to the FD inputs), and in the graphical output (there is a 'hole in the house').

We pause to reflect on a comparison between the Target Game and the activities with HOUSE. In the Target game the match between our priorities and the children's were clear and explicit; the task of interpreting feedback was straightforward (is the answer bigger or smaller than the one I'm aiming at?). In the HOUSE activities the underlying proportionality required by us was embedded and implicit: it needed to be *interpreted*. The key difference is that in the TARGET game, the appreciation of the feedback was much more closely bound into the articulation of the relationships involved: in the HOUSE situation, on the other hand, a perfectly reasonable interpretation of our request to build a bigger house in proportion could, as we shall see, bypass altogether the internal relationships of its structure.

Let us turn to an essential aspect of the system. Failure to adopt a multiplicative approach could be simply identified by the failure of the graphical output to resemble a house – the emergence of a 'hole' (see Figure 4.9).

We had hoped that students would see the problem and change the symbolic relations – but this turned out not always to be the case. Some students found the hole quite acceptable as it stood, adopting a pragmatic stance to this rather schoolish task. Others saw that the outcome was not a 'HOUSE' (houses shouldn't leak); but instead of reflecting on the relationships in their program, they debugged the outcome by 'homing in', closing the gap by adding turtle steps a few at a time after the figure had been drawn. This made the picture 'look OK' – and in terms of the students' agenda in relation to the activity, provided a reasonable solution. Of course, it may well have been that students who adopted this approach believed they were constructing a figure which *was* 'the same shape', interpreting sameness within everyday discourse; and it was in recognition of this possibility that we designed Activity 4.

4.4. *Towards a consistent approach*

Our new objective was to encourage students to pay attention to the internal structure of HOUSE, as it is only by focusing on the internal relationships between the walls of the house that one can guarantee that there would not be a hole. In the *Hole in the House* task, we presented to the class for discussion, a range of assessments of the proportionality of two houses; when the larger one had been constructed by adding and homing-in. The idea was that children had to decide individually which response they agreed with and justify their decision to the rest of the class.

This activity was off-computer, and it turned out to be significant: it brought all the different perspectives into the open so a consensus on approach could be reached. We surmise that it worked because of the previous computer work; crucially, students had developed a language in common, they had a shared medium with which to communicate. It is worth pausing to consider what sense students might have made of such an activity without prior computer experience. In the first place, without the computer, one strategy is very much as good as any other. That is, the student has little to assist in judging between strategies – other than by acceding to the whim of the teacher; the computer provided an alternative medium of explanation and validation, the 'neutral buffer' between teacher and student we remarked upon in Chapter 3. Second, what would there be to talk *about?* Each student had used HOUSE, and had been encouraged to discriminate the relationships between its constituent components: there was a basis for shared meaning between students and teacher, and between the students themselves. HOUSE *together with its properties* had become a friendly and familiar object.

We also encouraged the students to explore their preferred strategy more deeply, by initiating a game in which pairs of students were asked to produce 'enormous' houses in proportion to the original HOUSE, and record both their strategy and the numerical lengths of their figure. They then challenged another pair to find how their enormous house had been generated, giving their opponents only one length. Eventually, the two pairs came together to compare their findings and approaches.

The first stage of the game was designed so that students would construct a similar shape to HOUSE using a scale factor which was *not obvious* (i.e. not integral) – in order to present a challenge for their opponents. All pairs of students built 'enormous' houses with a multiplicative scale factor. They were not explicitly told to do so, but their behaviour was not altogether surprising given that their previous experiences on the computer had been moving in this direction. However, one result of observing movement to and from the discourses of Logo and pencil-and-paper mathematics was to underscore just how situation-specific the students' strategies were. We found that the *construction* of a HOUSE could call up a different strategy compared to the *interpretation* of the opponents' HOUSE.

The example of Toby and David is instructive. They constructed a larger HOUSE by multiplying each of the sides by 2.379. However when their opposing team (Paula and Jane) presented them with a larger house whose vertical side was 53.95 (the original length was 20), Toby and David edited their HOUSE procedure changing the first FD 20 to FD 53.95. At this point, Toby suggested that they add 8 to the next input to FD (the smaller HOUSE had a 'roof length' of 28, 8 more than 20), and they fixed on this additive strategy for the remaining lengths. Of course this left them with a problem for the base of the HOUSE, but despite their apparent understanding of proportional strategies in the construction part of the activity, they happily found the length (106) by homing-in – in fact, perhaps it is significant that they *wrote down* the length as 106.95 to give spurious accuracy to their result! They justified their result as follows: 'The side was 8 less than the vertical lines. We gest [guessed] the bottom line'.

The girls had constructed their house by multiplying by 2.6975, so that when the group came together, the four children were forced to resolve their differences. The two computer programs clearly set out the different approaches - the solution processes were available for inspection. Faced with the boys' solution, the girls immediately thought they had made a mistake and checked all their multiplications. Then, convinced that they were *not* in error, the girls discussed the two strategies and explained to the boys what they had done. David changed sides and supported the girls, agreeing that he and Toby had been inconsistent and pointed this out to his partner: ultimately the four came to a consensus. After this activity Toby made more progress – he began to attend to the consistency of his strategies in proportional situations, and at one point began quite noisily to proclaim in class discussion that 'the same had to be done to all the parts'!

5. QUANTITATIVE WINDOWS ON LEARNING

We have painted a picture of some of the *RatioWorld* activities, their rationale, and some insights derived from close observation of students interacting with them. We now face the question of what students learned, or more exactly, how their thinking about proportionality might have been reshaped during the activities. We wanted to find what kinds of connections the students made (not simply whether or not they made a specific connection) and how they were influenced by the nature and setting of the questions asked. In this section, we will discuss the results of the quantitative aspects of the study, and gather some quantitative support which can be set alongside our qualitative observations. We ask readers who maintain a healthy scepticism towards quantitative data in educational settings, to suspend judgement until we have had a chance to set them in their proper perspective.

Table 4.2. Total number of correct responses per item (maximum possible = 28)
* These items are included in the sub-category rugs/fraction in future analysis.

Context	Scale factor	Number of correct responses per item		
		Pre	Post	Delayed
	Function Integer	12	20	21
	Scalar Integer	12	15	17
	Scalar Integer	13	20	15
	Function Integer	17	20	18
PAINT	Fraction > 1	4	11	9
	Fraction < 1	5	10	7
	Fraction > 1	2	6	7
	Fraction < 1	0	3	5
	Scalar Integer	18	24	25
	Function Integer	9	17	16
	Scalar Integer	21	22	24
	Function Integer	14	19	24
	Scalar Integer/Ans Non Integer*	8	17	17
RUGS	Function Integer/Ans Non Integer*	7	17	11
	Fraction > 1	3	10	5
	Fraction < 1	4	8	5
	Fraction > 1	3	11	9
	Fraction < 1	3	9	4

5.1. *Gross trends*

We begin with the written tasks. All the tasks (pre-, post- and delayed post-) were marked, and we recorded correct and incorrect responses. One particular focus of interest was where the response coincided with an answer which would have been obtained if the *addition strategy* had been used (i.e., when the relationship was seen as additive rather than multiplicative). Next, we calculated totals and percentage totals of correct responses for the tasks overall and within the sub-categories defined by the *context* and *scale factor*. The integer questions were further classified by *operation,* that is, whether the operation used was function (between-measures) or scalar (within-measure).

Table 4.2 presents the raw data for the total number of correct responses per item, while Table 4.3 presents the data divided into the four sub-categories, *paint integer, paint fraction, rugs integer,* and *rugs fraction.*[14]

Tables 4.2 and 4.3, crude though they are, indicate that overall, the experience of working within the microworld setting led to some improvement in responses to the pencil-and-paper tasks (we will call this, for brevity, a 'microworld effect' – despite the misleading nature of the word 'effect'). Of course, we are not so naive as to believe that this tells us very much; but

Table 4.3. Total and percentage correct responses (n = 28) in
sub-categories: Paint Integer, Paint Fraction, Rugs Integer, Rugs Fraction
(% in italics)

	Number of correct responses		
Subcategory	**Pre**	**Post**	**Delayed**
Paint Integer	54.00	75.00	71.00
(no. of questions = 4)	48.21	67.00	63.39
Paint Fraction	11.00	30.00	28.00
(no. of questions = 4)	9.82	26.80	25.00
Paint	**65.00**	**105.00**	**99.00**
(no. of questions = 8)	29.02	46.90	44.20
Rugs Integer	62.00	82.00	89.00
(no. of questions = 4)	55.36	73.20	79.46
Rugs Fraction	28.00	72.00	51.00
(no. of questions = 6)	16.67	42.90	30.36
Rugs	**90.00**	**154.00**	**140.00**
(no. of questions = 10)	32.14	55.00	50.00
Overall total	**155.00**	**259.00**	**239.00**
	30.75	51.39	47.42

clearly there were some interactions between the very different settings; the
microworld and the written tasks. Representing the same data in diagram-
matic form (Figures 4.11 - 4.14) illustrates the trends in responses across the
three samplings; again they indicate that there is something worth explaining.

How consequential were these microworld effects? And what – if any
– were the interactions between the variables? To find out, we employed
an appropriate statistical test (the Repeated Measures Analysis of Variance)
which takes into account the hierarchical structuring of the observations
(subjects, time, context, scale factor and operator).[15]

Our first step was to check if there was any interaction between gender
and time – that is between the scores of boys and girls across the three
occasions pre-, post-, delayed-post.[16] We discovered that the trend over time
was the same for both boys and girls, implying that the effect of gender was
not conditional on time or *vice versa*. So it made sense to take the average
trend profile of all the students regardless of gender, as a basis for examining
differences over time. When we did so, we found that the changes over time of
these average profiles were significant within each of the four sub-categories.
In other words, statistically, it made sense to talk of a 'microworld effect'.

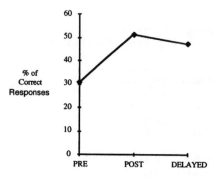

Figure 4.11. Overall % correct.

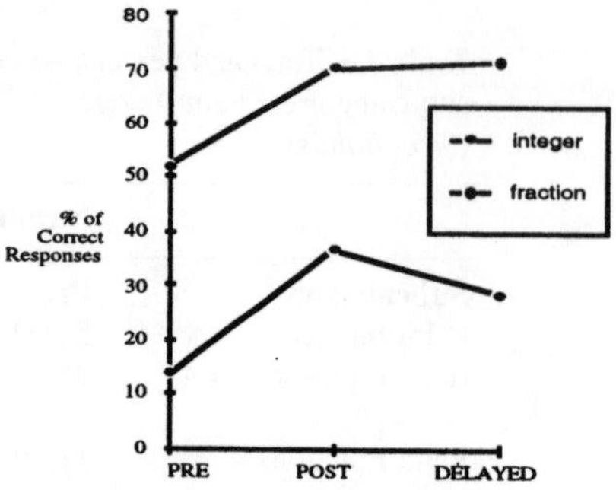

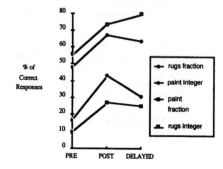

Figure 4.12. % correct classified by scale factor.

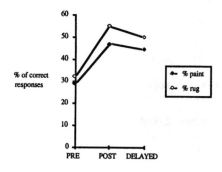

Figure 4.13. % correct classified by context.

Figure 4.14. % correct classified by context & scale factor.

What about other interactions? We repeated the analysis with respect to gender, time, context and scale factor, and found that there was a highly significant improvement in response rate,[17] but none of the interactions were significant. This reinforced our belief that we were justified in ignoring all other influences and eliminating the effects of all interactions in favour of an overall 'microworld effect', comparable for boys and girls and across scale factors and across contexts.

5.2. A step towards the individual

So much for gross trends: it looks as though some learning, connected with the *RatioWorld* activities, was taking place. We now took the first of a number of steps in which we harnessed statistical analysis to the task of trying to understand the behaviours of individuals, by hooking it up with the qualitative data we had gathered. Our first step in this direction aimed to find out which variables were responsible for generating correct and incorrect responses. In other words, we wanted to know what were the most influential determinants of students' answers, and how they interacted. This is a step in the direction of disaggregating the data, a move away from considering the consolidated effects of 'the microworld' on 'the students' – at best an oversimplification, and at worst, an insensitive generalisation. Our concern shifted to trying to pinpoint what may have been happening to individuals.[18]

There is a well-known technique for this kind of problem, that is to try to build a linear model of the responses. This is a relatively straightforward procedure.[19] In technical terms, it revealed that the simplest model that fitted the data was

$$context + time + scale\ factor + context \times scale\ factor.$$

In plain English, this means that i) the context of the task, ii) the scale factor and iii) the interaction between i) and ii), were influential in students' responses, as, of course, was the different occasions (time) that the written tasks were undertaken.

In order to further refine our model, we investigated (for integer questions only) whether the structure of the proportional relationship favouring a choice of function or scalar operator had an influence on response scores. This was analysed with respect to the different occasions *(time)* and in the two different contexts. Table 4.4 shows the raw data:

Table 4.4. Number and percentage of correct responses in Sub-categories Paint Function, Paint Scalar, Rugs Function, Rugs Scalar (% in italics)

Sub-category	Number of correct responses		
	Pre	**Post**	**Delayed**
Paint Function	29.00	40.00	39.00
(no. of questions = 2)	*51.79*	*71.43*	*69.64*
Paint Scalar	25.00	35.00	32.00
(no. of questions = 2)	*44.64*	*62.50*	*57.14*
Rugs Function	24.00	36.00	40.00
(no. of questions = 2)	*42.86*	*64.29*	*71.43*
Rugs Scalar	38.00	46.00	49.00
(no. of questions = 2)	*67.86*	*82.14*	*87.50*
Function	**53.00**	**76.00**	**79.00**
(no. of questions = 4)	*47.32*	*67.86*	*70.54*
Scalar	**63.00**	**81.00**	**81.00**
(no. of questions = 4)	*56.25*	*72.32*	*72.32*

This table suggests that there *was* some influence on responses, depending on whether the integer operation was scalar or function. However, we could not be sure as the number of items was relatively small. In such situations the statistically acceptable procedure is to use *a log-linear* model, which has the effect of smoothing out the distortions arising from small samples. When we did this, we found that the best-fit model was given by *context + time + context × operator*.[20]

In summary, both sets of quantitative analyses show that as well as the influence of time, student responses to the written tasks on proportionality were significantly influenced by context and scale factor. Further, they show that the influence of context cannot be separated either from the scale factor or from whether the sense of the questions gives preference to function or scalar operators.

5.3. *The baseline group*

We must briefly turn our attention to the baseline group. While we certainly were not able to manipulate our samples of children randomly, or to make any claims regarding the 'sameness' of the *RatioWorld* children and those in the baseline groups, we did at least hope to gauge whether any comparable changes in our microworld group's behaviours had occurred in the baseline groups, who had only engaged in the written tasks (on the three occasions).

In fact, looked at as a whole, the responses of the baseline group showed little change over time.[21] Applying the now familiar techniques with respect to the variables *time* (pre-, post-, delayed post-task), *group* (microworld, baseline) and *scale factor* (integer, fraction), we found significant effects.[22] These indicated that the difference between pre- and post-task scores depended on whether the group involved was *RatioWorld* or baseline as well as on the nature of the scale factor. Put another way, there was little change in scores for the baseline group over time in comparison to the change in scores for the *RatioWorld* group over the same period. From this we can make two deductions: first, it is extremely unlikely that the microworld effects were due simply to maturation; second, students learned rather little from the written tasks in isolation from other interactions.

5.4. *Individual student response profiles*

We were now in a position to make a final move into analysing individual students' responses. We knew that overall, the experience of working in the computer-based sessions had shifted the children's understandings in the direction of the mathematical – at least as measured by their performance on our written tasks. But these are aggregated statistics, telling us nothing about individuals, so it was impossible to relate them to the qualitative data we had collected. So we turned our attention to assessing the changes in individual students.[23]

One pattern stood out clearly from the data. At the pre-task stage, we found that students with little idea as to the mathematical meaning of 'in proportion' or who did not pick up context clues as to the structures underlying the question, adopted a 'pattern-spotting' approach – positing *any* pattern between numbers or figures as equally valid irrespective of the mathematical structure underlying the pattern. In the numerical paint context, the pattern-spotting in integer questions tended to be based on adding. But in the rugs context, it tended *not* to be adding; instead children's responses seemed to be influenced by visual clues gleaned from the context which led to the rejection of addition as a valid strategy. In fraction questions, adding was the predominant response for both contexts.

At one level, we can think of these data as illustrating students' tendency to be fearful of fractions and decimals – even with a calculator available. In the face of this sort of obstacle, they will tend to revert to the most familiar strategy – addition. This was certainly the case with the students we interviewed many of whom saw no conflict in adopting strategies which varied across contexts and scale factors. It is worth pointing out that from a theoretical point of view, we were meeting students who had a variety of techniques at their disposal, choosing according to features of the question – a rather sensible reaction but one which did not always accord with a mathematical perspective on the questions. We should also recognise that the pattern-spotting phenomenon can be attributed as much to the culture of U.K. school mathematics as to an individual child's mathematical competence (a point we discuss in more detail below).

We therefore wondered whether we could look more deeply at these individual findings using statistical tools. To do so, we made use of a technique which allowed us to undertake a more fine-grained analysis by looking at the combinations of responses of individual students *item-by-item*.[24]

At this level, we had some hope that the statistical analysis would highlight rather than blur the heterogeneity of the data, and focus attention on the process by which the children's understandings unfolded (an expectation which is rather contrary to the expectations of many statistical analyses we have encountered).

The analysis revealed that for 14 out of 18 specific items, there was a significant change in favour of a *RatioWorld* effect. The group-level analysis of the preceding section had indicated that context and scale factor were significant influences on student responses. We therefore integrated these two levels of findings by analysing individual student profiles of responses with respect to these same variables: paint integer, paint fraction, rugs integer and rugs fraction.

Now we could pinpoint significant divergences in the profiles of pupil responses. In particular:

- For *paint integer* questions, students with all the answers *incorrect* at pre-task produced most response changes at post-task. The majority of these changes represented a move from addition to a correct strategy.
- For *paint fraction* questions, the likelihood of a change from an incorrect to a correct strategy was approximately the same as the likelihood of change between incorrect strategies, whilst the probability of an incorrect response still predominated even at the post-task stage.
- For *rugs integer* questions, the changed student responses were no longer so confined to students who had all incorrect responses at pre-task. Students originally used a variety of strategies, and adding was not the dominant approach. Following the microworld experience there was a strong likelihood of change in the direction from an incorrect to a correct strategy.
- For *rugs fraction* questions there was a relatively high likelihood of change but in the main this was between incorrect strategies rather than from a correct to an incorrect strategy.

A summary of this analysis is given in Table 4.5 below:

Table 4.5. Pupil strategy changes between pre- and post-tasks.

	INTEGER	FRACTION
PAINT	Majority of changes from addition to a correct strategy	Likelihood of change from incorrect (both add and other) to correct strategy approximately equal to change between incorrect strategies (from add to other and vice versa)
RUGS	Likelihood of change from incorrect to correct strategy *not* dominant in pre-task)	Significant move away from addition, change equally likely from add to correct as add to other

Now we could return to our interview data, where we could look more generally at *strategy changes* which might account for these findings (as a methodological aside, we point to the interplay between quantitative and qualitative analysis). Here we discerned a clear shift between pre- and post-responses:

- from a perceptual to an analytic approach;
- from inconsistent or pattern-spotting approaches towards an awareness of the underlying mathematical structures and the consequent *need* for a consistent approach; and
- from implicit strategies towards a conscious, explicit formulation of the operation which had to be performed.

Although these latter two shifts were influenced by context, it seems likely that some pupils had developed a feel for proportional situations, building for the first time a connection between these situations and a multiplicative strategy. A schematic overview of our findings is presented in Table 4.6:

Table 4.6. Changes in student strategies.

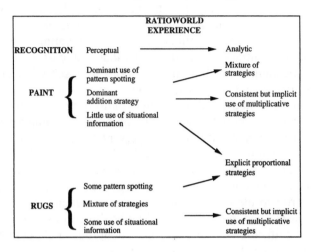

6. REFLECTIONS

Our investigation with *RatioWorld* employed a mix of quantitative and qualitative data: the former drawing our attention to connections and worthwhile questions, the latter offering some hope for explanations and tentative answers. In fact, this methodological eclecticism will characterise other studies we present later in the book, and we will discuss it in more detail in Chapter 6. We are now in a position to consider the implications of the study.

We begin with an important insight provided by Nunes and her colleagues:

> ... given the existence of two sources of knowledge about proportion – activities in and out of school – with two different approaches to problem-solving and no systematic attempt to establish a relationship between them, it is likely that the two practices will form separate systems of arithmetic, just as oral and written calculations do.
>
> (Nunes, et al., 1993, p. 122)

Our findings support Nunes's assertion that meanings derived in and out of school are 'separate systems'. More importantly, they suggest ways that these sources of meaning can be connected. Although we had no data concerning extra-school practices, our discovery that students' responses were contingent on question context, indicates that extra-school meanings are far from homogeneous. The meanings drawn from the rug and paint scenarios differed significantly: with the essentially numerical setting of paint encouraging a 'pattern spotting' approach, while in rugs, there was a greater tendency towards using situational information. One way of interpreting this finding is that the context was 'reinvented' in the classroom: as we indicated earlier, the paint problems may have evoked familiar classroom strategies in which tables of numbers are routinely used to spot numerical patterns, whereas the same may not have been true for sets of geometrical figures.

So it might be a mistake to think of rugs as generating only 'everyday' meanings: on the contrary, it might well be that – at least within the school setting – the meanings brought to the questions were school meanings, even if they were insufficient or inadequate ones. As we saw, students used the 'default' addition strategy, and were reluctant to bring situational meanings into play. In the rugs context, there was more use of situational information: and we cannot rule out that the representational form of the problem – consisting of pictures – made it more likely that meanings derived from common-sense or everyday settings were mobilised, and that this in turn led to a rejection of the incorrect addition strategy which characterised the paint setting. But prior to the microworld experience, students were inconsistent in their responses even in one context. Maybe this inconsistency arose from the multiple signification of the questions: about rugs or about mathematics? Perhaps at this point, these influences worked against each other – particularly if the context called up a particular type of operator, either function or scalar. Meanings must come from somewhere; in the absence of any obvious candidate they can be drawn from a variety of situations. For rugs, there was a graphical object which could provide a source of meanings: for paint, there was no equivalent source – but there were tables of numbers, and there were school-meanings to attach to them.

What of the trajectory of students' thinking? Here too, we have seen just how complex the situation was, and how sensitive it was to the context of the question, the operation involved, and the presence of non-integer scale factors. It is clear that the *RatioWorld* experience changed strategies across the board: equally clear is that this was only sometimes from an incorrect to a correct (proportional) strategy. But here a clear difference emerges: while none of the settings tended to produce strategies which could be characterised as proportional before the activities, it was the integer questions which demonstrated most chance of producing 'correct' responses afterwards.

Our interpretation of this finding is tentative, but it will point towards a fruitful line of analysis in what follows. It seems that whatever the approach (adding or otherwise), integer questions were most likely to end up with proportional approaches. What can we make of this? Perhaps we might conclude that what was taken from these situations was *not* to do proportional questions or perform proportional operations, but something else: the need to establish a *relationship* between the elements of the problem, rather than focus on the objects themselves, and to do this in a *consistent* way. Perhaps we are seeing the 'something *other* than algebra or geometry' which, as we saw in the previous chapter, Papert proposes as the fundamental learning outcome of microworld settings.

Suppose this conjecture is true, suppose that playing with relationships in the form of lines of programming code, encouraged the students to see that focusing on relationships was a way to bring new meaning to the questions posed. In that case, we might take it for granted that this would be the case *only* where the relationships between objects were evident, manipulable, visible. We know from existing studies, that this is generally only true for whole numbers, and very much *not* the case for fractions or decimals. Paradoxically, the failure of students to develop towards proportional strategies in non-integer cases, acts as confirmation of the influence of the *RatioWorld* activities.

A simplistic corollary of the observation that there are two different systems, is to think of the everyday as a concrete situation, full of meanings, and the mathematical as an abstract, meaningless domain. In this scenario, the best that can be achieved is to try to hold in view the concrete (the 'referent') to any abstraction, so that meanings can at least be derived from somewhere. This is the (often implicit) approach which underpins the 'real world' pedagogy beloved of text book writers. It establishes an artificial relationship between in-school and outside-school by assuming that no such relationship exists.

Suppose instead that in the course of their *RatioWorld* activities students were developing new meanings, making connections not only between the world of lengths and quantities, but seeing that meanings can be derived from relationships as well. Suppose further that students came to view the manipulation and construction of objects and relationships in *RatioWorld* as a new situation from which meanings could be drawn. In that case, we have what appears to be a paradoxical situation: have we observed the construction and interpretation of lines of programming code – an 'abstract' activity, surely – used as a resource for conferring meaning to questions about paint and rugs – a 'concrete' activity?

If so, it would appear that through working in the microworld, the tools, though formal, became convivial – the algebra of the relationships meaningful. It may be in this way that situational information and mathematical relations can remain connected while the latter is generalised where it seems to be appropriate. In fact we see here the possibility that formal and algebraic can be used as a mechanism for creating meaning, rather than erecting a barrier to it. In the next chapter, we elaborate this further. To do so, we will need to look more broadly at how learners make meanings from activities within settings such as *RatioWorld*.

NOTES

1. This study was part of the 'Microworlds Project', funded by the Economic and Social Research Council, under grant number C00232364. The investigators were Celia Hoyles, Richard Noss and Rosamund Sutherland. The remainder of this chapter is based on the previously unpublished final report of the project, Volume III: Hoyles, Noss and Sutherland (1991b).
2. Cited in Nunes et al (1993), p.78.
3. A full review of the research in this area is beyond the scope of this chapter. It is available in for example in Tournaire (1985), Behr, Harel, Post and Lesh (1992), Carpenter, Fennema and Romberg (1993) and Harel, G., & Confrey, J. (1994).
4. CSMS: Concepts in Secondary Mathematics and Science (see Hart, 1981).
5. We are aware that the notion of knowledge domain is problematic (see, for example, Lave, 1988), and we have some sympathy with Lave's view in global terms. Nevertheless, we need some way of referring to an accepted and identifiable group of local mathematical ideas. For further analysis of these different types of situation from a cognitive point of view, see Nunes & Bryant (1996).
6. The other two are: *product of measures*, and *multiple proportions*. See Vergnaud (1983).
7. So that if an item is designated, 'function integer', it means that if a between measures operator is chosen it would be integral.
8. This is a more sensible designation than 'rational', as Logo handles non-integer values as decimals.
9. This formulation owes much to Wilensky (1991) whose work we have considered in Chapter 2.
10. Ourselves and Rosamund Sutherland.
11. Students had about 25 lessons in Logo, 35 minutes each, and had undertaken a Logo project on tessellations over a period of about 20 hours.
12. The classes in the baseline group had different mathematics 'attainment' levels than the experimental class, and were in no sense a 'control' group. However, they had no teaching of ratio and proportion during the period of research.
13. This game was based on a similar (calculator-based) one in Swan (1983).
14. For the purposes of this analysis the two rugs questions which involved division by an integer giving an answer which was not an integer were included in the sub-category rugs fraction. Thus the categories *integer* and *fraction* predominantly describe the nature of the scale factor but also include, in the case of rugs fraction, a description of the nature of the solution. The decision regarding this classification was taken after an analysis of the pre-test responses to these items.
15. The traditional cross-classification ANOVA, that is factorial design, was inapplicable given that the occasions (pre-, post- and delayed post-tasks) and the outcome measures were nested within subjects. The significance test employed was the F test.
16. Had there been such an interaction it would be misleading to compare gender means averaged over time, or time means averaged over gender groups. For each of the four sub-categorise (paint integer, paint fraction, rugs integer, rugs fraction) analysis of variance tables were drawn up. We discovered that the boys (n = 16) scored consistently (and significantly) higher than the girls (n = 12) but there was no *gender* × *time* interaction effect.

17. The results of the RMANOVA for all the scores with respect to *gender*, *time*, *context*, *scale factor* and the interactions between these variables gave an overall significant improvement (F = 13.90, p < 0.00001, number of degrees of freedom = 2.52).

18. This, or rather the failure to do this, is the reason why many have turned their back on statistical analysis altogether. We sympathise with the rationale of those who do so, but as we shall see, this position is unnecessarily restrictive.

19. We used GLIM, (the General Linear Interactive Modelling Program). The attempt to build a comprehensive model, starting with all possible variables and interactions, and successively removing components of the model which did not significantly contribute to the fit, was as follows:

Model	χ^2	Degrees of freedom	
c + t + s + cxs + cxt + sxt	3.7114	2	A good fit
c + t + s + cxt + sx t	5.8546	3	No significant change (no cxs interaction)
c + t + s + cxt	9.1139	5	No significant change (no sxt interaction)
c + t + s + sxt	5.8955	5	No significant change (no cxt interaction)
c + t + s + cxs	7.1790	6	The simplest model

c = context ; t = time; s= scale factor

20. The Log linear model (forwards procedure) of pupil responses to integer questions with variables context, time, and operator was as follows:

Model	χ^2	Degrees of freedom	
Null Model	52.321	11	Significant (not a good fit)
o	50.44	10	No significant change (*operator* is not significant)
o + c	43.33	9	Significant change (*context* is significant)
o + c + t	19.822	7	Significant change (*time* is significant)
o + c+ t + cxt	19.351	5	No significant change (*context × time* interaction is not significant
o + c + t + oxt	17.362	3	No significant change (*operator × time* interaction is not significant)
o + c + t + cxt + oxt + cxo	0.103	2	Significant change (*context × operator* interaction is significant)

c = context ; t = time; o= operator

21. It is also fair to say that we are suspicious of the authenticity of some of the results, particularly for one set where the student responses strongly suggested a lack of motivation for the task (a great many omissions). We see this as a general problem of research requiring written test data.

22. $F = 3.63$, number of degrees of freedom = 2,132 at the 5% level.
23. Pooling the data of post- and delayed-post task results.
24. McNamara's related groups chi-squared test. Effects referred to in the subsequent paragraph were significant ($p < 0.05$).

CHAPTER 5

WEBS AND SITUATED ABSTRACTIONS

1. REVIEWING THE FOUNDATIONS

In Chapter 2 we laid the foundation stones of our work and saw the problematisation of dichotomies such as formal/informal, concrete/abstract, contextualised/decontextualised as our primary challenge. We questioned the extent to which there was any justification in thinking of mathematical ideas in terms of a global hierarchy, and in particular, the arrangement of such a hierarchy along a concrete/abstract dimension. Our contention was that this misperception arose from restricting attention to what individuals learn, to the interaction between the individual and knowledge domain. By emphasising questions of meanings and their construction we hoped we would avoid conflict between mathematics as an object of study and mathematics in use. In particular, we sketched the framework for a theory of mathematical meaning which would transcend the purely situated view of cognition, while simultaneously respecting the epistemology of mathematics. There were secondary concerns: we were interested in how the resources of the setting structure and are structured by participants' mathematical activities and the meanings they construct. Following Vygotsky's ideas of tool use, we focused on the dialectical relationship of action and thought which synthesised the idea of thinking tools, communicative tools and (real) tools, and this led us to focus, in Chapters 3 and 4, on the ways in which computational tools could be exploited by learners in the construction of mathematical meanings.

Our analysis so far has pointed towards a key concern: to develop a more conscious appreciation of mathematical abstracting as a process which builds upon layers of intuitions and meanings. We have begun to illustrate how the creation of meanings can be understood as a dialectic between actions and formalisation, how computer-based environments can open windows for learners on new intellectual connections and for teachers and researchers on the ways mathematical meanings are constructed. We have argued that there are fruitful analogies to be drawn between interacting in microworlds such as *RatioWorld*, and engagement in practical, everyday, activity: the analogy draws attention to the intellectual tools routinely used as a way to represent and communicate ideas about actions. And we have located as a central question for learning and teaching, how to assist students in building on their own meanings in ways which converge towards mathematical ways of knowing.

We want to put forward a case for learning as the construction of a *web* of connections – between classes of problems, mathematical objects and relationships, 'real' entities and personal situation-specific experiences. Recog-

105

nising the importance for developing meaning of building connections is not a new idea – indeed finding good analogies and metaphorical images is at the heart of good teaching. This approach has characterised recent work on new mathematics curricula: for example, Paul Goldenberg and Al Cuoco have chosen activities for their Connected Geometry programme, largely because of their potential to help students link up different aspects of mathematics, different realisations of concepts in a variety of situations (Cuoco, Goldenberg, & Mark, 1995; Goldenberg, Cuoco, & Mark, 1993). Based on series of case studies of learners struggling with probabilistic concepts, Wilensky (1993) has put forward the notion of 'connected mathematics' within a constructionist framework, and has advocated the need to connect personal intuitions with formal mathematical definitions. He makes a powerful case for designing learning environments with multiple interconnectivities, in which mathematical knowledge can be 'concretized' rather than built on an edifice of brittle procedures.

This 'connectionist' trend will serve as a powerful metaphor for building our theory of mathematical learning, and synthesise the different paradigms introduced in Chapter 2. We note in passing that researchers in this tradition cast the computer in a central role as a means to build connections through experimentation and construction. Perhaps it is no coincidence that a number of workers in the field are choosing the connection metaphor to characterise their approach, and are seeing the computer as a central element to operationalise it. After all, the model of intelligence bubbling up from networks of connections is one with which the Artificial Intelligence community have worked for some time, and it is hardly surprising that some in the educational community – especially that section concerned with computational media – have adopted the metaphors of AI and Cognitive Science with whom it shares a desire to understand the construction of knowledge.[1]

Focusing on technology draws attention to epistemology: for new technologies – *all* technologies – inevitably alter how knowledge is constructed and what it means to any individual. This is as true for the computer as it is for the pencil, but the newness of the computer forces our recognition of the fact. There is no such thing as unmediated description: knowledge acquired through new tools is new knowledge, MicroworldMathematics is new mathematics. Researching how students exploit autoexpressive computational settings to communicate, (re-)present and explain, not only provides descriptions of how individual students can express mathematical ideas, but can provide more general clues to the processes involved in learning, how knowledge is modified in the direction of mathematisation. In our now familiar metaphor, it provides a window both on student schemes and their transformation: a *re*-vision of mathematical knowledge.

We set out in this chapter to develop some insight on two questions:
  i. how the resources of computational environments support students' mathematical activities, and

ii. the nature of the abstraction process within computational settings.

2. WEBS OF MEANING

It is clear that we need to focus on *mediations*, on that which stands between the individual and social learner, and the 'knowledge' which he or she is supposed to learn. From an educational point of view, it comes as no surprise that the teacher can play a critical role in the learning process, no less in the computational setting than in the traditional one. This point is worth reiterating, in case we give the impression that we see the learner as a lone constructor of mathematical knowledge: in fact, we have more to say about teachers in later chapters.

If teaching is to figure alongside learning, we have as much to gain from Vygotsky as we do from Piaget. Vygotsky introduced his celebrated zone of proximal development as a way of transcending the evaluation of learning in terms of tasks that a student can currently master, to include a vision of what he or she has the potential to master. The ZPD is:

... the distance between the actual developmental level as determined by independent problem-solving and the level of potential to development as determined through problem-solving through adult guidance or in collaboration with more capable peers.

(Vygotsky, 1978, p. 86)

Vygotsky argues that a child who might show no signs of 'readiness' on his or her own can be encouraged to enter and be successful in a new task with the assistance of a more able adult or peer. There are clear pedagogical implications of this view, most notably developed by Wood, Bruner and Ross (1979), who pointed to the central role in didactical theory of *scaffolding* – graduated assistance provided by an adult which offers just the right level of support so that a child can voyage successfully into his/her zone of proximal development. More generally, cognitive psychologists working within this tradition have shown that learning can be facilitated by providing help in developing an appropriate notation and conceptual framework for a new or complex domain, allowing the learner to explore that domain extensively. Learner participation is gradually increased – according to the needs and learning pace of the individual – and the support is gradually *faded*. Thus scaffolding/fading is a metaphor to describe a carefully designed form of instruction where interventions are contingent on a broad instructional framework.

Some while ago, we proposed that the metaphor of scaffolding might usefully be extended to computational settings (Hoyles & Noss, 1987). We asked ourselves in what ways a computer might play the role normally ascribed to a human tutor? There were, however, a number of difficulties associated with the wholesale import of the scaffolding notion. First, there is a connotation in scaffolding of a structure erected around the learner by an external agency. External agencies may abound (classrooms, textbooks, curricula), but our concern is centred on the ways in which learners structure their own learning, as well as on the ways in which the setting structures it. Second, while the idea of a 'zone' is a useful metaphor, it suggests the idea of a bounded territory, on which scaffolding can support some building – but it leaves open how the zone defined and how its limits are delineated. Third, we are unhappy with an idea of fading which suggests that the computer is providing support which ultimately has to be removed – leaving what? We would like to borrow from the notion of scaffolding the idea of a support system, but we would prefer a metaphor which does justice to the fluidity and flexibility of computational settings under the control of the learner,[2] and which recognises the value of computationally-based understandings in their own right.

The distinctive features of this support system we want to capture are:
- it is under the learner's control;
- it is available to signal possible user paths rather than point towards a unique, directed goal;
- the structure of local support available at any time is a product of the learners' current understandings as well as the understandings built by others into it;
- the global support structure understood by the user at any time emerges from connections which are forged *in use* by the user.

Where the metaphors of the nineteen-seventies were still structured by the practices of architecture and building, it will come as no surprise that we have sought a new metaphor grounded in newer technologies: we will borrow from the ubiquitous idea of the *Web*. The etymology of the (World Wide) Web is obvious to anyone who has ever used it; an expanding, interconnected, network of sites, each one built by an individual (or group of individuals), yet drawing on the resources available at the time of building. Like the web of mathematical ideas, the Web (we will use a capital to denote the electronic network), is too complex to understand globally – but local connections are relatively accessible. At the same time, one way – perhaps the only way – to gain an overview of the Web is to develop for oneself a local collection of familiar connections, and build from there outwards along lines of one's own interests and obsessions.[3] The idea of *webbing* is meant to convey the presence of a structure that learners can draw upon *and reconstruct* for support – in ways that they choose as appropriate for their struggle to construct meaning for some mathematics.

The analogy we intend should be clear enough. We clarify it further by outlining three ways in which webbing extends the notion of scaffolding, although we do not claim that these extensions are obvious from our choice of metaphor. First, the literature on scaffolding is mainly concerned with the support of skills in practice such as basket weaving, while we intend webbing to apply as well to the learning of conceptual fields such as mathematics.

While an expert-learner form of support may be appropriate for the appreciation of a well-defined goal – to be able, say, to produce a basket independently – it seems less obvious that it could function effectively for the learning of a complex knowledge domain such as mathematics. For this kind of robust and deep learning, we borrow the constructivist tenet that effective learning involves the cognising subject in building his or her own intellectual structures, and taking – more or less autonomously – what is supportive from the ambient pedagogical setting, rather than 'receiving' what is given.

Second, the literature on scaffolding indicates that it is most often seen as universal; that is, having voyaged onto the scaffolding erected in the zone of proximal development, the child is assumed to have 'developed' in a sense which is independent of the context in which the learning takes place. In contrast, we view webbing as domain contingent and even within a knowledge domain such as mathematics, it focuses attention on the influence of the setting and the symbol system within which the ideas are expressed.

The third way in which webbing extends the idea of scaffolding is in its ceding of control over the support structures to the learner. The Brunerian formulation seems to imply a fixed and immutable goal, an implication that there is an acceptable form of, say, mathematical progress independent of children's constructions. Vygotskian theory largely centres on the goal and motive of the child's problem-solving – in fact, Activity Theory is founded on the notion that a goal, even if not set by the child, should be shared by the child. Yet is it possible that learners retain their autonomy when the specification of the goal is in the hands of a teacher? If we take webbing as the fundamental motor for the construction of meaning, we can envisage a situation where we respect the 'epistemological freshness of children's invention and its implications for the revision of expert knowledge' (Confrey, 1995, p. 205), while at the same time forge connections with the teacher's mathematical intentions. Webbing is a way to reconcile Piagetian and Vygotskian approaches – there are connections built into the structure of the environment, and even signposts which assist in navigation: yet the structures discovered, and the signposts followed (and ignored) are largely in the hands of the learner.[4]

Achieving the delicate balance amongst connections needed to develop mathematically, is by no means simple. When the connections to the teacher's goal are dominant, students may simply imitate a procedure or try to guess what the teacher wants; when they are not made at all, the mathematical agenda might be bypassed (although if we recognise diversity and reject the notion of a unique line of development, the situation is not so brittle).

We now provide some specific instances of the webbing idea. Our intention is to provide glimpses sketched with only a cursory introduction, drawn from a variety of sources to illustrate its generality. All the examples involve the computer. This is not meant to imply that webbing only takes place during interactions with computers. Rather, it is in this context that connections can more readily be built and, equally important, the processes of webbing, its shaping by and within the medium, are more accessible to the observer. In what follows, we largely focus on elementary mathematical ideas to introduce the theoretical ideas in the simplest of scenarios. In Chapter 10, we aim to show the continuity of this approach with more advanced mathematical structures.

2.1. *Flagging*

Our first example of webbing is based on a small study we undertook some years ago.[5]. We had been working with a group of seven students aged between 13 and 14 years with considerable Logo experience (around 120 hours over four years). We set out to observe a small number of students playing with Logo-models of parallelograms.

  We gave the students the following procedure and encouraged them to play with it and then to use it in the construction of a tiling pattern:

```
TO SHAPE :SIDE1 :SIDE2
FD :SIDE1
RT 40
FD :SIDE2
RT 140
FD :SIDE1
RT 40
FD :SIDE2
RT 140
END
```

Our hope was that by undertaking a carefully designed series of tasks, the students would notice how the relationships in the procedure's code were necessary and sufficient for SHAPE to produce a parallelogram. That is, we wanted their attention to be drawn to the facts that:

  • the inputs to the procedure (SIDE1 and SIDE2) represented the lengths of the two sides of the figure;
  • SIDE1 and SIDE2 were each called twice and alternately;
  • the inputs to RT represented the turtle turn between the drawing of the sides;
  • the sum of the two turtle turns (if in the same direction) between adjacent sides was constant and equal to 180.

After the initial exploration, we gave the students a series of on- and off-computer tasks to assist us in probing student conceptions in a more systematic way.

While trying to generalise SHAPE, one student, Amy, wrote the following procedure:

```
TO SHAPE1 :SIDE1 :SIDE2 :MOVE
FD :SIDE1
RT :MOVE
FD :SIDE2
RT :MOVE
FD :SIDE1
RT :MOVE
FD :SIDE2
RT :MOVE
END
```

This procedure opens a window on which relationships Amy had discriminated: she seemed to see opposite sides as equal, but a casual observer might assume that Amy believed that all the turtle turns to form a parallelogram must be equal as she gave the same name, MOVE, to all the inputs for angle. However, it appears that this was not the case. Amy immediately edited her procedure *without running it* and added an extra input MOVE1:

```
TO SHAPE1 :SIDE1 :SIDE2 :MOVE :MOVE1
FD :SIDE1
RT :MOVE
FD :SIDE2
RT :MOVE1
FD :SIDE1
RT :MOVE
FD :SIDE2
RT :MOVE1
END
```

So Amy's intention in her first version of SHAPE1 seemed to have been to *flag* the overall program structure – to put down its shape and the positioning of the different types of commands – knowing that later she could worry about the details. The fact that her final program used the 'adding-on' strategy for inputs – a different input for everything that varies – served to indicate that she may not yet have been explicitly aware of the necessary relationship between the angles of a parallelogram. Nonetheless, in her use of SHAPE1, Amy *only chose turns which added up to 180*. She knew how to use the program as a tool but had yet to connect the relationship embodied in the medium with a geometrical principle.

Now this is quite an illustrative, though simple, example of a more widespread phenomenon. Amy used the Logo program as a way of expressing the relationships as she saw them, and as they changed for her. The way in which she flagged these relationships is characteristic of what anyone does using a medium to place a marker for an idea, one which they will flesh out later – an outline sketch, a note etc. The partial relationships which Amy embedded for herself in the program nicely catch the way in which the computational setting provided a web of structures which she could control and exploit at a particular moment, shaping the resources of the technology to suit her purpose. Her actions were deeply connected to the visual feedback; but at this stage, they may have been ahead of her means to articulate the relationships. This gap – between her partial understandings and more general recognition of the relationships – was part of the web of ideas indexed with the computer. The critical features of the problem were marked, modified, and finally built into a mathematical structure, but one that could be modified later. We surmise that if Amy had more experience with her program and shifted her attention to the numerical size of its inputs while keeping contact with the screen images, she would be able to move smoothly to changing her two inputs to one, by recognising the relationship between them.

It is worth pointing to the relationship between Amy's flagging of the relationships involved, and her emerging sense of what characterises a parallelogram. We do not know precisely what Amy thought, of course, but at some stage it must have been something like:

> It's a procedure with four inputs two for the sides and two for the angles. You do forward and turn four times, with the lengths and the angles alternating. When you call the procedure the sum of the two turns has to be 180.

Embedded in this description is the language of turtles. And here we begin to locate a disjunction between the computational and the mathematical. The symbolism of a program certainly involves flagging relationships in programming code; it allows the learner to focus on the essence of what is varying and what is invariant, but not in a way that necessarily matches standard mathematics.

2.2. *Adjusting*

We have already briefly introduced LESLI (see Chapter 4), one of the computational objects (alongside HOUSE) which formed part of the *RatioWorld* (see Figure 5.1).

We gave students the following task:

> Produce 'families' of different *types* of pinfigures all *in proportion* by editing the LESLI procedure (for example, a proportional family of pinfigures with small heads).

Many students identified with the pinfigure, played with it, made it dance, run, sit down and play. Our intention was eventually to focus attention on the idea that only a certain type of modification – multiplicative – would produce

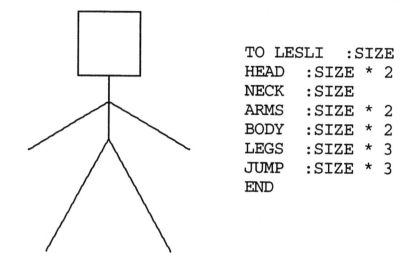

```
TO LESLI  :SIZE
HEAD  :SIZE * 2
NECK  :SIZE
ARMS  :SIZE * 2
BODY  :SIZE * 2
LEGS  :SIZE * 3
JUMP  :SIZE * 3
END
```

Figure 5.1. Pinfigure LESLI.

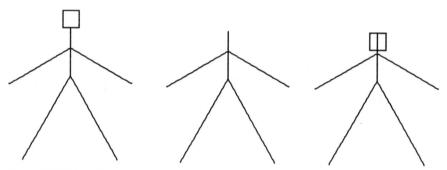

Figure 5.2. Using subtraction to construct pinfigures with small heads: here the input to HEAD in LESLI was (:SIZE – 20).

classes of figures that were in proportion, since strong visual feedback would draw attention to the non-proportionality of figures constructed by other methods. Not surprisingly, a wide range of strategies were used, many of them resulting in deformed LESLI's: in particular, those who used 'additive'[6] strategies to, say, shrink LESLI, soon found that it generated figures with non-existent heads, or with heads which disappeared into the body (see Figure 5.2).

We rather hoped that feedback of this sort would encourage pupils to reflect on the way the procedure LESLI had been constructed and to discriminate the necessary nature of the relationships between the constituent subprocedures in order to obtain sets of pinfigures in proportion. Yet, like the 'Hole in the HOUSE' task of Chapter 4, many students did not see this.

There are two complementary issues. The first is that inviting students to adjust a template of our (i.e., the researcher or teacher's) making, does not guarantee that the adjustment will satisfy our expectations. This is hardly surprising as it is a simple manifestation of the play paradox: who owns the activity and its tools? But it is also endemic to the activity: the underlying proportionality of the LESLI activities was embedded in the code. To be sure it could be read from the code, but that assumes that all students are equally inducted into the rules of reading and deciphering the meanings within it. The relationship between the code and the graphical output is *not* straightforward; moreover, what actually *is* webbed from the embedded relationships depends on a dialectic between the structures which form those relationships, and the local mental structures which the student currently possesses. In this case, simply adjusting the template provided by the task might not *automatically* produce the insight we hoped for.

On the other hand, by *using* LESLI in different ways, interacting closely with the program, some students did notice the intra-LESLI relationships. For example, by a process of elimination, many students came to notice and articulate something important which we might re-express as:

A family of LESLIs will be in proportion, if all inputs to any subprocedures of LESLI are :SIZE multiplied by some number – with no adding or subtracting.

The crucial point about this observation, and it is one to which we shall return, is that this formulation is both correct (within the medium of the programming code) and constrained, as it does not explicitly articulate the general relationship involved – which might be something like *Proportionality implies a multiplicative relationship* or its converse. It relies on the Logo code as webbing for the partial articulation of the relationship.

It could be argued that the relationship as it is expressed above, entirely misses the 'essential' point. A tempting interpretation is that the rule is merely 'instrumental', that the student noticed a 'trick' which works, rather than 'really' gaining 'insight' into the relationships. The quotation marks in the previous sentence are there for good reasons: the precise meaning is heavily dependent on how far the syntactical rule is connected to the student's actions and visual interpretations, and on where the boundary is drawn around mathematics. We cannot know if the students who adjusted the LESLI template thought they were 'doing mathematics'. The decision about the boundary's position is negotiable; it is a question of the meanings which the setting evokes. We can agree that attention was focused (and articulated) on both the structure of objects and the relationships between them, even if these meanings might not have have been expressed conventionally.

2.3. *Sketching*

Cleo and Musha (age 14) were given the two flags shown in Figure 5.3 where one was the image of the other after a reflection[7] and their task was to find the 'mirror line' – the line of symmetry.

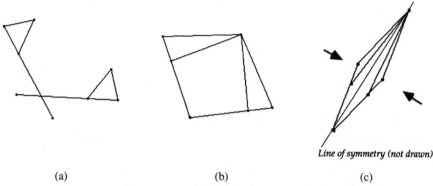

Line of symmetry (not drawn)

(a)   (b)   (c)

Figure 5.3. Snapshots of a strategy for locating the line of symmetry of the two flags. (a) illustrates the two flags. In (b) two corresponding pairs of points have been dragged to coincide. In (c) the students are beginning to drag each point until it coincided with its image. The line of symmetry has been drawn for clarification purposes: it 'appears' only when all pairs of points are dragged together.

The two girls did not know the constructions necessary, and so, not unnaturally, dragged[8] the basic points for some time, playing within Cabri to try to generate some clues. Slowly the activity became more focused and they started to drag the first flag about the screen a little more systematically, noticing the effects on its image. It was clear that they had a strong visual sense of where the mirror must be, a sense honed by their use of the medium – until after a short while, they could run their finger along its position on the screen with some certainty. However, this intuitive certainty gave no help as to *how* to construct the mirror: after all, construction would require a language which would allow them to express their intuition by doing more than simply pointing to the screen.

Suddenly Cleo picked up the mouse, and started to drag together the topmost and bottom-most points of the flag and its image (Figure 5.3 (b)). 'That's it!' they both exclaimed, drawing the line joining these intersections with their fingers on the screen. They then went further and dragged all the equivalent points together actually 'constructing' the required line (Figure 5.3 (c) illustrates the process).

Now this is a correct strategy for finding – even, in an unusual sense, constructing – a mirror line. The girls were completely confident at this point and could articulate their general method, their Cabri theorem, verbally:

> The mirror line is what you see on the screen if you drag points and their reflections together.

Note how this observation is generalisable to all reflections; yet it is only meaningful – most importantly, *constructable* – using the webbing which operates within this particular computational environment, the mathematical resources and connections built into the system. The solution to the problem is *sketched* within the medium by a physical manipulation controlled by perception – seeing that the points and their images coincided – yet satisfying a mathematical relationship. Now it could be argued that this Cabri theorem is just the same as the observation that the mirror line is where to fold a piece of paper to make corresponding points coincide.[9] However, we argue that it can offer more, as illustrated in the girls' subsequent activity. First, Cleo and Musha sketched the mirror in a new way by 'fitting' a line with two points on top of their 'squashed' flags. This meant that they then could dispense with their particular piece of the web. They knew they had not yet finished the task, but they realised that this strategy would allow them the freedom to move the flag (and its image) with the mirror line in place, so they could attend to the relationship between the three objects on the screen – the mirror, the flag and its image. Cleo and Musha were not merely looking for a visual pattern within a simulation devised by someone else: the mirror had been constructed by them. This process of construction clearly influenced their interpretations of the computer feedback as they dragged the flag; their focus was on 'equivalent' pairs of points, and this eventually led them to see that they could construct their mirror line by joining the midpoints. How could they do this?

A Cabri midpoint is itself an abstraction. It wraps up much of the process of construction into a set of actions – opening the appropriate menu item and clicking on two points – which become part of the web of ideas and actions embedded into the medium. The computer 'knows' what kinds of objects it 'needs' for a midpoint – a line segment or a pair of points. So some of the mathematical essence of the midpoint comes ready-wrapped; it is a tool to construct objects and *at the same time* an object which encapsulates a relationship. This duality is, as we have seen, at the root of mathematical activity, and it is precisely this duality which many have considered to be so problematic.

This example of using the webbing of the medium to sketch a solution strategy raises again the question of connections to standard mathematics. How far do Cleo and Musha regard their midpoint strategy as general and applicable beyond the medium? *We* see the generality beyond the tool use; but why should they differentiate this tool use from their previous 'squashing' method – the meanings of the two approaches were clearly intertwined, which accounted for the smooth transition they effected from one to the other. How might connections be made to other approaches beyond Cabri?

Consider the question of constructing a midpoint. Paper and pencil construction would demand compasses (or, of course, measurement!) and it is clear that these actions are only 'the same' as the Cabri construction from a particular perspective – one in which the role of the midpoint is already understood. The differences – and therefore, the difficulties – of seeing the mathematical abstraction in the computational setting, should not blind us to the fact that there are similarities as well. Nevertheless, we see here one indication of what it might mean to generalise from one setting to another: to become aware explicitly of the relationships wrapped into the setting, to notice precisely what elements of the computational web are interacting with one's current state of understanding. Using the web as a tool is a necessary but not sufficient condition for this awareness to emerge.

2.4. *Patterning*

Two 12 year-old girls were attempting to generate a set of polygon patterns using a spreadsheet (see Figure 5.4).[10] Penny was in control of events at this point.

The pair explored various ways to construct the triangle numbers – simply spotting a pattern, trying it out for one triangle number and, if it worked, trying it again on another. They recognised that a rule for the generation of one triangle number had to apply to all (a non-trivial observation) but made no attempt to control their conjectures by drawing the polygons, and simply embarked on a numerical search.[11] They homed in on the solution, trying to construct the triangle numbers as a linear function of the *position* of a number. This did not work, so they tried looking at linear functions of the *previous term*. This led them to see that the relationship involved *both* parameters, at which point, Penny exclaimed:

> Ah I've got it! It's *this* one add *this* one, *that* one add *that* one equals *that* one, *that* one add *that* one equals *that* one.

This utterance is deeply embedded within the computational setting, which renders it difficult to interpret from the text alone. *In situ*, its meaning was perfectly clear, as it was accompanied by gestures, the actions of pointing to cells of the spreadsheet. In fact, these gestures captured the relationship between two independent variables and the value of the triangle number, not explicitly but rather as exemplified by the cells. Thus the *pattern* in the gestures framed by the spreadsheet array was exploited by the students as the webbing they needed to build the values of the triangle numbers, and to express the relation that any triangle number could be constructed from the previous triangle number together with its position.

When they turned to the generation of the square numbers, the students made a smooth transition to a new set of connections; they already knew (from their school experience) how to square numbers and simply applied this method within their new language of interaction with the spreadsheet.

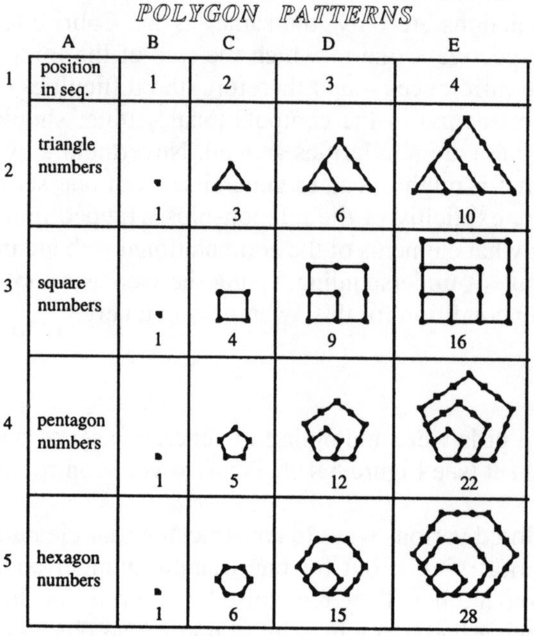

- **Generate the triangle numbers on a spreadsheet**
- **Generate the other polygon numbers**
- **Investigate different ways of generating the sequences**

Figure 5.4. Polygon numbers.

Constructing the pentagon numbers, however, was a different story. The two girls spent a considerable time trying out different patterns unsuccessfully. At one point they became aware that a rule for pentagon numbers might have to take into account both relevant triangle numbers and position. Penny said:

> Add ... oh hang on ... the position add the [triangle] number before, times it by however far down you are, times it by 4 'cos you're 4 down.

Again, it was the synthesis of the framework of the spreadsheet cells and the pattern of gestures which provided the webbing for articulating the relationship. In order to communicate more accurately to her partner, Penny had begun to supplement her gestures by a natural language description of the spreadsheet array, as indicated in the italicised phrases below. This had the effect of pushing her towards generality on the basis of the specific example: a pentagon number is *3 times the triangle number before + the position.* She knew that the '3' was an expression of *however far down you are,* so she could then generate the hexagon numbers, simply by changing the '3' to '4' and thus generalise to all polygon numbers giving:

A polygon number = (However far down you are) × (triangle number before) + position.

There is clearly a gap between this formulation, and its re-expression in mathematical language, which would be something like:

The n-th term of a k-gon number sequence is given by $P_{k,n} = (k-2)T_{n-1} + n$ where T_i is the i-th triangle number.

What can we say specifically about the way the medium webbed these generalisations? The most obvious point concerns the way in which some of the abstraction process came 'ready made' to the student: the casual use of the *copy right* feature masks a wealth of mathematical abstraction, just as the midpoint construction did in the previous Cabri example. We might say that precisely because this aspect is so powerful it veils the underlying mathematical generalisation implicit in its action: in effect, it appears as if the generalisation is done for the learner. Not so: the software webs this step, but it leaves the critical step of seeing the particular case as a prototype of the general in the hands of the learner. So in this instance, the webbing involves stepping on to the structure erected (by the copy key) around the formula; and it is invoked by the student explicitly recognising the pattern: what works for two cases actually works in general. A web is a resource; an environment which is webbed affords – but does not guarantee – resource-*ful* learning.

3. DOMAINS OF SITUATED ABSTRACTION

The previous section illustrated four ways in which (rather special) computational media structured and were structured by students' emerging mathematical ideas – this is the crux of the webbing idea. We now turn to a more thorny issue: what, in general, can students take from such activities? How, in other words, can we rescue the student from the mathematical limbo in which we may have left them?

On the one hand, we could characterise the kinds of interactions described in the previous section as irrevocably rooted within the computational medium, unconnected with official mathematics. On the other hand, some (including, in earlier times, ourselves), have been willing to blur the distinction between computational and mathematical, even to the extent of suggesting that they are effectively the same. Our position now is that links between the two are not obvious, but they *can* be forged. This process of building connections turns out to be central to making mathematical meanings, with the added advantage of challenging the simple dichotomy – mathematics or not mathematics.

Our exploration necessitates that we resolve the hierarchical dilemma implicit in the notion of abstraction. The traditional way to think about this is to think of mathematical objects as necessarily involving the abs-traction (pulling away) of objects into formal systems of relationships and opera-tions, divorced from any points of reference in the real, 'unmathematised' world. Viewed from this perspective, a setting is mathematical to the extent that the person has *already* abstracted the relevant relationships into a formal (sub)system. To the extent that the learner has yet to make this formal, abstract reflection on his or her actions, the situation can not be seen as mathematical *per se*.

How might we encourage individuals to abstract objects and relationships in a way that we would recognise as mathematical? One way would be to organise situations that *we* might recognise as mathematical, and use them as settings in which to develop mathematical ways of thinking. Now there is plenty of scope for tautology here: mathematics instruction often presupposes that students already possess mathematical ways of seeing and thinking, when it is precisely this way of seeing that we are trying to encourage. As Thompson (1994) points out in the context of the ontologies of space and time which are often assumed by teachers:

> ...to assume students will have any understanding of 'distance divided by time' we must assume that they already possess a mature conception of speed as quantified motion. This places us in an odd position of teaching to students something which we must assume they already fully understand if they are to make sense of our instruction.
>
> (*ibid.* p. 225)

In one version of the traditional view, the mechanism by which ideas and experiences become interiorized is, as we have already mentioned, 'reflective abstraction'. Here is the Piagetian motor which drives the process by which ideas become loosened from the hold of immediate experience, and become assimilated into higher levels of thinking and understanding. But reflective abstraction can only provide part of the story. For in its pure form, it ignores the situatedness of mathematical cognition. As we outlined in Chapter 2, we have to find some way of taking account of the boundedness of learning, of the problematic nature of transfer. In particular, while it may adequately describe the state achieved by those who have reflectively abstracted – to the satisfaction of the mathematician – it does not begin to describe how people come to abstract, and offers only general help in understanding the relationship between what is abstracted and how such abstraction occurs.

The notion of situated cognition is not unproblematic either. Its strength is that it challenges the separation of what is learnt from how it is learnt: '...situa-tions might be said to co-produce knowledge through activity' (Seely-Brown,

Collins, & Duguid, 1989, p. 32). So performance, say on mathematical tasks, has to be viewed as a complex construction – being a product of activity, context, history and culture. The difficulty is that if we accept that both language and actions mediate knowledge and actions, and that schemes of actions are moulded by the tools available and the context of the activity, then every cognitive act must be viewed as a specific response to a specific set of circumstances. Mathematical knowledge becomes bound into a setting. Does this make any sense from an epistemological point of view? Can mathematics be simply a collection of specific cases? Surely the field of mathematics is about general forms, comprehensive ways of seeing, universal truths, abstractions which transcend contexts? If mathematics is fragmented into LogoMathematics, PencilMathematics, CabriMathematics and so on, we will need to know something of how linkages can be forged between understandings and learning constructed in separate settings. How, for example, can Amy's appreciation of a parallelogram be connected to its more usual definition? How can a Cabri midpoint come to be seen as 'the same' as a Euclidean midpoint? As a corollary, we will also have to problematise the relative status of the different articulations, perhaps even question the historical assumption that PencilMathematics represents the highest pinnacle of mathematical abstraction.

The situated view of cognition, however, is not a monolithic theory. While there is general agreement on the ways in which coming to know is substantially more local and situation-specific than was previously thought, the mechanisms for learning continue to be the subject of debate. As we discussed in Chapter 2, there continue to be attempts to understand traditional Piagetian findings in situated terms. Edith Ackermann's proposal for a 'differential' approach, illustrates the point:

> Stage theory stresses the progressive decontextualisation of knowledge during ontogeny, while the differential approach provides a more situated perspective on knowledge construction – a framework for studying how ideas get formed and transformed when expressed through different media, when actualised in particular contexts, when worked out by individual minds.

(Ackermann, 1991, p. 380)

Ackermann argues for a compromise between differentiation and universality claiming that immersion *in* as well as emergence *from* embeddedness are both essential processes in the development of understanding of the self, and of the world. Her ideas move us from a focus on performance and an analysis of how performance is structured by setting, to a recognition that although knowledge is constantly constructed and reconstructed through experience, this same experience also shapes and reforms a global and theoretical perspective. It is just such 'global and theoretical' perspectives which are so important for mathematics; we shall exploit this insight as a way out of the mathematical cul-de-sac which a strictly situated view poses for mathematical learning.

The examples of the preceding section neither encompass a 'new vantage point defined by a set of axioms' (Tall, 1991b), nor have the stamp of approval granted, for example, by a mathematical proof. Yet they are replete with articulations which represent *something* – they are abstractions in one sense, they certainly – following the etymology of the word – *pull away from* the real referent and they focus on relationships rather than merely on objects. What we have learned so far, is that activity in a setting has influences beyond that setting, shaping global and theoretical perspectives. The distinction between situated and universal is too coarse: we need to add to the *general* situatedness of cognition, the *particular* situatedness of abstraction.

3.1. *Situated abstraction*

We intend by the term *situated abstraction* to describe how learners construct mathematical ideas by drawing on the webbing of a particular setting which, in turn, shapes the way the ideas are expressed. Of course, one could argue that all abstractions are situated, all expression mediated – so the adjective *situated* is superfluous. But we want to focus attention on the specificities of the situation, and in particular, on the linguistic and conceptual resources available for expressing mathematically within them, as well as the ways in which expressions within a situation can point beyond the boundaries of that situation.

Official mathematics – PencilMathematics – is frequently viewed as decontextualised, unsituated. In fact, it constitutes a distinct domain in which the official language of communication and expression is, for the most part, algebraic. So one intention behind the notion of situated abstraction, is to take a step nearer viewing the official expression of mathematics as just one of a possible myriad of generalisations expressable in widely different forms. In other words, we want to decouple the definition of what constitutes mathematics from the specificities of the language in which it is expressed.

Our substantive point concerns meaning. Abstracting can be seen as a way of layering meanings on each other, connecting between ways of knowing and seeing, rather than as a way of replacing one kind of meaning with another. Our point is that this process can happen especially appropriately in computational worlds. In such settings (but not *only* in such settings) meanings are simultaneously preserved and extended in the constructive process. What can and does happen – as we saw earlier in this chapter – is that these meanings become reshaped as learners exploit the available tools to move the focus of their attention onto new objects and relationships.

In our early thinking about situated abstraction (see, for example, Noss & Hoyles, 1992) we emphasised its meaning as an entity, a thing – and Papert (1992) understandably suggested that this notion was closely allied to his 'object to think with'. While we are happy with this connection, we want to underline the dualistic nature of the idea. Situated abstraction is a process as well as an object: abstracting *in situ*, abstracting in a domain. We shall also refer to a domain of situated abstraction as a 'place' (in intellectual and physical space) where abstractions of this kind arise naturally. This will allow us to use the word 'abstracting' as a verb: situated abstraction is not a thing on its own, it is simultaneously an 'articulation', a 'statement', a (re)thinking-in-progress.

Before we provide some examples, it is instructive to consider a related idea. In her analysis of human machine communication, Suchman (1987) describes the relationship between *action* and a-priori *plans* (and retrospective accounts). The thrust of her argument is that actions in the course of social practices are not the realisation of plans, but are *situated actions:* actions which depend in essential ways on their material and social circumstances (Suchman, 1987). Plans are not identified with action, and although they may be resources for action, they do not, in any strong sense, determine what is done. They may, however, *represent* what is done.

The traditional account of purposeful action assumes that there is a set of factors and alternatives which are knowable in advance, and that an individual's course of action simply consists of choosing between them. Suchman proposes that plans are only weakly associated with actions, and that they are merely the reconstruction – either before or after the event – of what has been, or is to be done. There is a disjunction between our actions and our understanding and articulation of them:

> ...rather than direct situated action, rationality anticipates action before the fact, and reconstructs it afterwards.
>
> (*ibid.*, p. 53)

Despite her focus on planning, Suchman is clearly interested in the more general question of the relationship between rationality and action, between how people describe what they have done (or will do), and what they do. She points out that the intellectual tradition of the cognitive sciences presupposes a clear delineation between theory and practice, between abstract and particular instance. She characterises the cognitivist view as follows:

> An adequate account of any phenomenon, according to this tradition, is a formal theory that represents just those aspects of the phenomenon that are true regardless of particular circumstances. This relation of abstract structures to particular instances is exemplified in the relation of plans to situated actions. Plans are taken to be either formal structures that control situated actions, or abstractions over instances of situated action, the instances serving to fill in the abstract structure on particular occasions....
>
> (*ibid.*, p. 178)

It is here that Suchman's ideas converge with our own.[12] For the traditional view of mathematics is that of an overaching structure which can be read into, or is absent from people's activities. It is commonplace to say *this* is a mathematical activity, or *these* are the mathematical ideas embedded within it. This abuse of language masks a deeper, more fundamental assumption, that actions and activities are mathematical to the extent that the actor is *thinking* mathematically, that the structures and relationships underpinning the activity are *known* to the actor.[13] Insofar as they are known, he or she is doing mathematics. Insofar as they are not known, he or she is shopping or selling.

Mathematics is more than action in the sense of activity-with-objects. It is activity-with-relationships, it is a virtual reality of abstractions in the form of relationships, justifications and generalisations. We cannot avoid the idea of drawing-away-from action in order to make 'real' or 'understood' that which before was imaginary and unknown. Our use of the term situated abstraction is designed to underscore the idea that the condition of 'being abstract' does not come ready-made, either *a priori* or *post hoc*. Rather, abstraction is a process which develops in activity, which – like all activity – is situated. Like plans and actions, abstraction is not necessarily a guide for and determinant of mathematical activity, it is a *resource* for activity. This constitutes the criterion for recognising a situated abstraction; it provides new meanings, new resources for the learner.

In many settings, mathematical invariants underpin actions, and learners come to exploit these as a functional way to achieve their goals. But these invariants may be unexpressed, implicit in the action, and observable by someone else – someone versed in mathematical ways of thinking. By contrast, we have seen how within computational environments, some of these invariants are rooted in action and articulated – quasi-mathematically – in the operational terms of the available tools.[14] Our claim is that students 'breathe life' into sections of the computational web. By this means theoretical reflection on their processes of construction is promoted and knowledge is gradually reconstructed towards a more conscious appreciation of form and content.

A straightforward example is the Cabri reflection task we saw earlier. How does the construction – and as far as we can tell the appreciation – of the symmetry of the graphical display differ from that achieved after playing with, say, 'Brush Mirrors' in MacPaint? Primarily, the distinction lies in the construction process: MacPaint would have achieved the same result (in fact, with considerably less trouble), but at the expense of decreasing drastically the chances of the mathematical structure being discriminated – as a domain of abstraction, MacPaint is considerably less mathematically rich than Cabri! This is a general point: mathematics requires a language – there must be a preservation of the balance of action and description.[15] As Laborde (1993a) puts it:

> Such programs [Cabri, Logo etc.] differ from drawing tools like MacPaint in which the process of construction of the drawing involves only action and does not require a description. In this kind of software, construction

tasks are no longer to produce a drawing but to produce a description resulting in a drawing.

(*ibid.* p. 54)

Our central tenet, therefore, is that a computational environment can be a particularly fruitful domain of situated abstraction, a setting where we may see the externalised face of mathematical objects and relationships through the window of the software and its accompanying linguistic and notational structures.

3.2. *Webs and abstractions*

Let us paint the most general picture, before we put some flesh on the bones of our theory. Our two notions, *webbing* and *situated abstraction*, display an essential complementarity. Our explanatory framework has attempted to capture the ways in which meanings are constructed by action on virtual objects and relationships. Within a computational environment, some at least of these objects and relationships become real for the learner (we are using 'real' here to mean something other than simply ontologically existent – perhaps *meaningful* or broadly connected are better descriptions): learners web their own knowledge and understandings by actions within the microworld, and simultaneously articulate fragments of that knowledge encapsulated in computational objects and relationships – abstracting *within*, not *away from*, the situation. In computational environments, there can be an explicit appreciation of the form of generalised relations within them (the relational invariants) while the functionality and semantics of these invariants – their meanings – is preserved and extended by the learner.

To what extent is this kind of meaning-making unique to autoexpressive computational media? To answer this question, we begin by observing that mathematics *can* be done without a computer! It *is* possible to preserve and extend the meanings of relational invariants, and to make 'concrete' for oneself, the apparently most 'abstract' of objects. How else do mathematicians talk of n-dimensional manifolds as old friends, or the cosets of an infinite group as tangible and manipulable? In building a linguistic framework of mathematical expression, mathematicians create for themselves meanings on which new expression, new relationships can be constructed.

We do not propose to usurp the mathematician's prerogative, only to extend it to a wider audience. It is possible to design settings involving the requirement to express generality as a result of manipulating computational objects, and where a productive synergy of webbing and situated abstraction can be generated. For in an autoexpressive computational medium, the expression of the relational invariants is not a matter of choice: *it is the simplest way to get things done*. In this case, the action/linguistic framework of the medium supports the creation of meaning: and it is here that computational media differ substantially from inert environments.

Recall Cleo and Musha trying to find the mirror line with Cabri. We have seen how they appeared to be 'certain' of where the line should be, but unable to articulate how the position could be derived before they experimented with the software. What *did* they know initially? They knew where the line should be, in the sense that it either looked right or it didn't. What they didn't 'know' was a mathematical model of their intuitive knowledge, a piece of formal, systematised knowledge which would help them to construct it. As we saw, the medium provided some kind of bridge between these two states: the representations the students built elaborated and illuminated their knowledge structures and simultaneously provided a window to and a route beyond these structures.

But this is not the end of the story. How do we interpret these student constructions? How do we bridge the evident gap between them and constructions that somehow transcend the medium? Construction implies an explicit appreciation of the relationships that have to be respected within any situation, a mathematical model of the situation (how else do you know what to focus on, and what to ignore?). The key insight is that parts of this model are built into the fabric of the medium, thus shaping the types of actions that are possible: they do not *only* exist in the mind of the learner. The level of what can be thought about, talked about, is notched up a rung or two: and abstractions can be made which are situated within the existing model, expressed through 'statements' (mouse clicks, pieces of programs etc.) which are already expressions of mathematical abstractions. It is the web of relationships and objects offered by the computer that can act both as a support for developing new meanings, *and* as a means for transcending that support. For by manipulating objects and articulating the relationships between them, a dual action/notational framework is developed which can begin to be a *new* resource, one which is not so dependent on the medium for its expression.

Here is the analogue of 'fading' in the webbing metaphor: not the removal of the support system in the form of the resources of the setting, not the progressive suppression of contextual 'props' to 'reveal' the mathematical knowledge required for problem solution; rather its connection with other settings, other notational systems, and other meanings.

3.3. *Connecting drawing to geometry*

We now provide an example, taken from Jones (1993), which illustrates how a connection can be made between a computational medium and official mathematics. Two undergraduate students, Tommy and Charlotte were working in Cabri Geometry; their task was to construct a circle that is tangent to two lines and has a given point P as its point of contact with one of the lines.

The students began by creating a circle, choosing a centre somewhere between the two lines with P as a radial point, and then adjusting its size

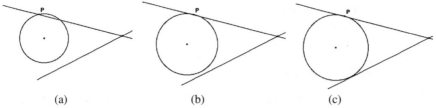

| (a) | (b) | (c) |

Figure 5.5. Tommy and Charlotte judge the position of the centre and size of the circle by eye. In a), the circle does not have either line as tangent, although it does, by construction, go through P. In (b), the upper line is (approximately) a tangent. In (c) both lines are judged tangential.

and position purely by eye so that the circle appeared tangential to the two intersecting lines as in Figure 5.5.

Tommy and Charlotte were aware that they had not really solved the problem:

> T: So that's done like that. But how can we .. is there a way we can be sure that we've got .. I mean we just lined it up by sight didn't we? There must be a way of being absolutely sure.

But although their 'product' was little more than could be achieved with a drawing program, manipulating the circle with the available tools (crucially, the 'drag' mode) and seeing it in position (note that the circle was constructed to go through P) provided Tommy and Charlotte with just the support they needed to make a first step in the construction process:

> C: ... Well, the tangent is perpendicular to the line of radius, isn't it?
> T: Right so we need to put in ...
> C: we can put a line in ...
> T: a radius in ...
> C: and then measure the angle.
> T: Right.
> C: Can't we do actually a perpendicular line?

Following Charlotte's observation, the pair constructed a line perpendicular to the line containing the point P. They then tried to fit a circle in between the two given lines, and hit upon the strategy of coping with one variable at a time. They used their perpendicular line as a way of fixing their circle, and created a circle with centre on the perpendicular line through P. They then moved this centre along the perpendicular until they had only one point of intersection with both lines. But again they knew they had not quite finished:

T: But, again, ... umm ... is that just ... that could just be ... a fluke. We put it there and it might be that the resolution of the screen is such ...

C: OK, take the circle away then ...

C: ... and see if we can draw two perpendicular lines from the two ... from P, a perpendicular line from P.

They explore for a while longer and then Charlotte had an idea:

C: So we could draw 90° perpendicular lines from here, from this line here [the lower of the two intersecting lines], perpendicular lines, and somewhere, where it hits, the same distance, presumably ... is it going to be the same distance? ... the radius is going to be the same? ... to both lines ... probably, not sure about that ... do you see what I'm saying?

Charlotte wanted to draw a line perpendicular to the lower line and move it around to try to think how to construct the centre. Not a bad idea: using this approach, there would be no alternative but to experiment, as the point of contact on the lower line was, of course, unknown. But this turned out to be a virtual experiment only: Charlotte's words were enough to trigger in her partner the idea for a construction:

T: Yes ... Ah! Now would the centre of the circle lie ... I'm just thinking something slightly different now, because I'm just trying to think, there must be a way of securing the centre accurately ... and I'm thinking ... does the centre of the circle ... sit on the bisector of the angle that's made by those two lines?

C: Oh right .. could do, yes, well, draw that one, draw that one, and then we want a line perpendicular from where, if you draw that one like you say, right, have a go at that one, and that will cut that line at some point and then we draw a perpendicular line from that point to this line [the lower of the two intersecting lines] and that should be the centre of our circle and we'll see if this circle will work.

T: OK, so ...

They bisected the angle between the two intersecting lines, and had solved the problem.

Imagine trying to adopt this strategy without the computer. Seeing these connections by exploiting the web of relationships embedded in Cabri allowed the students to find a solution by eye, and provided them with a pathway which indicated the insufficiency of their intermediate answers as well as pointing to a resolution. It also provided a route which could be progressively abandoned as an alternative strategy presented itself. The example nicely illustrates the dialectic between the ways in which the resources of the software structured the students' developing ideas (P was fixed on the circle, the perpendicular line was an obvious and straightforward extension, the location of the perceptual tangent was relatively accessible) and at the same time, how the students' activities structured the software's resources (the choice of the various menu items was far from arbitrary).

This example differs from previous extracts in this chapter, in that the students ultimately employed geometry to solve the problem. To achieve this involved developing a Cabri approach which triggered a connection to geometry: the solution itself was *not* situated within the medium. This connection was only possible as the students (unlike those in the earlier examples) were already familiar with a range of geometrical constructions. Their experimentation with Cabri made them aware of the constraints on the centre of the circle which called up the idea of angle bisector – *but with a new functionality*: not as a tool to bisect angles but as a means of specifying a locus of points equidistant from two lines.

In passing, it is worth noticing how we, as researchers, are granted a way of appreciating the geometrical intuitions that the students had – because they could model them by positioning parts of the construction by eye. How easy it would be to jump to the conclusion that drawing was simply 'wrong'! Equally, notice the difficulty – perhaps the impossibility – of using this 'wrong' strategy as a starting point for a more correct one, without the web of mathematical objects and relations built into the computational system.

4. TOWARDS A THEORY OF MATHEMATICAL MEANING

Meaning...is the primary human motion, irreducible. It cannot be divorced from the body, from social experience, or from the very survival of the organism. Meaning depends on someone who recognises you. Not meaning, by definition, is utterly lonely. Well-fed, warm and free of disease, you may still perish if you cannot 'mean'.

(Kegan, 1982, p.19)

Meaning in mathematics, like concreteness (Wilensky, 1991), is a relationship between person and knowledge. Meaning can be derived from intra-mathematical sources, from the discourse in which mathematics is expressed: yet we have argued that this does not imply the necessity to sever form from content. On the contrary, our argument so far has emphasised that mathematics can be considered just like any other sign system – mediated by symbolisations which are cultural phenomena, which structure and are structured by social life. A few are recognised by others immersed in the same sign system, and through this system, they recognise others. Meaning is established. For most, the discourse of the sign system is too far removed from that of everyday discourse for recognition mediated by mathematical discourse routinely to occur: it looks as though meaning is *only* established intra-mathematically. However, viewed through the window of the computer screen, we have seen that this is mistaken: mathematical meaning can be derived from being recognised by a computer.[16]

As we have seen, this recognition involves a dialectical interaction between tool and language in the course of interaction. It is inevitable, therefore, that different settings will generate different meanings, that knowledge will be, as diSessa has put it 'in pieces' (see diSessa, 1988b). On the other hand, these pieces of knowledge can be robust precisely because they are built within the boundaries of a domain of abstraction where meanings are locally webbed by the medium and not wholly dependent on hierarchies of intra-mathematical meanings which are dependent on pre-existing mental constructions.

At the same time, while we might want to consider a mathematical 'essence' shared by, say, the expression of a relationship in a spreadsheet and a formal expression in algebraic terms, they are certainly not the same for the child, perhaps not the same at all. We find it helpful to think of algebra, like any microworld, as a domain of situated abstraction, where the linguistic framework consists of a language carefully constructed in order to facilitate the articulation of (virtual) relationships and relationships between relationships. Viewed in this light, our task is to find ways to assist in this process, means to support the forging of connections across domains. How this might be done is a matter for research. Our hypothesis is that in this process, both learners *and* tools will come to develop new meanings.

A full discussion of this problem will have to wait until the final chapter, where we can mobilise a clearer picture of the diverse ways in which meanings can be established. But we can already point to one corollary: we can dismantle the hierarchical relation which is routinely perceived between mathematical and other discourses. The idea of situated abstraction allows us to revalue meaning itself: to challenge the notion that mathematics achieves its power by suppression rather than extension of meanings, and that the very existence of meaning mitigates against generality. We return to the contention of Nunes et al.:

> ...the fact that specific information is contained in representations in street mathematics is not a drawback. It is specific information that allows subjects to control for the meaning and reasonableness of their answers in problem situations. ... Thus, representation of the particulars of a situation does not imply that the subject is restricted to understanding exact situations. There is ample evidence for flexibility and generalisability of the pragmatic schemas of street mathematics.

> (*ibid.* p. 147).

Extending the argument from the street to the microworld setting using our new constructs, there is every reason to assert that situated abstractions are by no means inflexible and incapable of generalisation beyond the boundaries of the situation in which they are derived.

It is fundamental to the Vygotskian view that understanding how people acquire scientific concepts requires a perspective which also includes an analysis of children's spontaneous concepts. As Saxe (1991) puts it:

> For Vygotsky, children construct 'spontaneous concepts' from the bottom-up: ... Such spontaneous concepts are rich in meaning for children, but they are local and not linked with one another in general systems of interrelated understandings ... [in contrast] children construct 'scientific concepts' from the top-down ... Initially, such concepts are abstract but empty; while the child may learn the 'syntax' of the relations of a network of concepts, these concepts have little apparent relation to the child's spontaneous concepts. ... In their interaction, spontaneous concepts enrich scientific concepts with meaning and scientific concepts offer generality to the development of spontaneous concepts.

> (Saxe, 1991, pp. 12–13)

For the moment, let us accept Saxe (and Vygotsky's) sharp division of concepts into scientific and spontaneous. Then we might identify a domain of situated abstraction (with or without the computer) as a setting in which the interaction between the two sorts of understandings are brought together and where the dialectical relationship outlined by Saxe is played out.

This is an appealing resolution of the problem. The difficulty with it is that it seems to suggest that meaning emerges *only* from spontaneous concepts, that the 'role' of scientific concepts is to 'offer generality' but not meaning. This strict division between spontaneous and scientific is, of course, merely another face of the dichotomies which abound in attempts to understand learning: concrete/abstract, informal/formal, explicit/implicit. A key distinction has always been the role of language. For example, Polyani states that tacit knowing can *never* be explicitly verbalised – it is intuitive, non-intellectual: '...we can be aware of certain things in a way that is quite different from focusing our attention on them' (Polyani, 1968, p.31). He draws a contrast between two kinds of awareness that he suggests are mutually exclusive; a from-awareness (tacit) and a focal awareness (explicit): 'in the from-awareness of a thing we see a meaning which is *wiped out* by focusing our attention *on* the thing' (emphasis in original).

Now it is not our intention to propose that such a distinction is completely invalid, any more than we think that scientific discourse is coterminous with everyday or street discourse. What we reject is the mutual exclusivity of the two kinds of meanings, the 'wiping out' of meaning in transition from tacit to explicit awareness. Rather, we see meanings – far from being wiped out – as being reshaped and recreated in action. Our description of webbing as a motor for situated abstraction blurs any strict division between what is implicitly and explicitly known, precisely because language is necessary for action with things as well as for focusing attention on the thing itself. Meanings can be established in the course of systematisation, as much as by reference to 'context' or action.

Finally, we return to the play paradox. At first sight, we seem to be arguing that we can design computational worlds which are finely tuned to particular mathematical wavelengths, and which guarantee that involvement within certain activities constructs just the right set of mental object. Of course, nothing is so simple. The essence of exploration is the discovery of the unexpected and the play paradox is designed to capture its pedagogical consequences. However, if we consider the idea of situated abstraction as not involving abstraction in a Platonic sense but rather the naming and objectifying of some structures or relationships within a setting, we can propose a resolution to the play paradox: namely that the system carries with it elements of what is to be appreciated, it arranges activity so as to web solution strategies. The structures which form that web of meanings are, of course, determined by pedagogical means: they are not arbitrary. There is human intervention, mathematical labour, dormant within the system, and the learner has to breathe life into it. It is this interaction which points to a resolution of the play paradox by converging the expectations and goals of students and teachers.

It is important to recognise that situated abstractions are shaped by mathematical cultures, not merely by technologies. How can children appropriate mathematical ideas in ways which give sustenance to their creativity? One answer, of course, has already been given: by building the learning environment appropriately. Others remain to be discussed which take account of the environmental and social elements which make up the learning culture. It is to these elements that we now turn.

NOTES

1. A seminal work in this regard is Marvin Minsky's (1986) *Society of Mind.*
2. We also want to reject any connotation which might suggest that the computer is cast in the role of imitating a human tutor.
3. We have resisted writing 'surfed', as we have no guarantee that next year's netspeak will coincide with this year's.
4. Since writing this chapter we have belatedly read the work of Scardemelia & Bereiter (1991) much of which resonates with our position in that they design environments (not specifically for mathematics) where students can assume a higher level of agency in learning. Turning particularly to mathematics, the French didacticians (as we noted in Chapter 3) describe these types of situation as *adidactical.*
5. Derived from Hoyles and Noss (1987).
6. Actually, in this case, a 'subtractive' one – simply 'taking away' something from each component.
7. This episode emerged within a study conducted with Lulu Healy and Reinhardt Hoelzl.

8. 'Dragging' is a technical term which describes the way in which any basic point in a construction can be 'picked up' by the mouse and moved about the screen.

9. We are grateful to Colette Laborde for raising this question and provoking us to further elaborate the role of the medium.

10. For more details of the episode, see Hoyles, Healy and Sutherland (1991a).

11. These two observations – the tendency to strive for an empirical match and the neglect of the visual in a spreadsheet environment – have been noted in a more recent study; see Hoyles & Healy (1996).

12. We must confess to having read Suchman's book after we had begun to formulate our own idea of situated abstraction.

13. An obvious example is the assumption in some psychological and sociological research that anyone using numbers in any form, is necessarily engaged in mathematical activity.

14. We recognise that it would be helpful in another publication to undertake a detailed comparison of our idea of situated abstraction with Gérard Vergnaud's notions of a 'theorem in action' and 'conceptual field' (Vergnaud, 1982; 1990).

15. There are, of course, significant differences between the 'language' of a programming environment, and the 'language' of Cabri, which includes gestural and iconic aspects. These differences are not insignificant pedagogically, but we do not discuss them here.

16. We hope that all that has gone before alerts the reader to the power, as well as the dangers, of casting the computer in an anthropomorphic role.

CHAPTER 6

BEYOND THE INDIVIDUAL LEARNER

> The most neglected existence theorem in mathe-
> matics is the existence of people.
>
> P.C. Hammer. *cited in* Pimm (1987, p. xvii)

1. EXTENDING THE WEB

So far, our notion of webbing has emphasised meaning-making as the preserve of the isolated individual and teacher, and a domain of abstraction comprising a student in seclusion with a suitable means of expression. In this chapter we redress this imbalance, and take steps to reinstate the corollaries of Hammer's existence theorem by addressing some of the problematic issues surrounding the collaborative construction of mathematical knowledge. While maintaining a respect for the individuality of student contributions, we take the first of several steps towards appreciating more broadly the culture in which students construct mathematical sense, and the diverse meanings brought to bear on any problem situation.

Taking into account the social dimension of mathematical meaning-making hugely increases the complexity of our project, and this will necessarily involve further elaboration of our methodological approach. At the core of this chapter, an investigation which exploited a mixture of quantitative and qualitative techniques as introduced in Chapter 4 is elaborated.

The task is all the more complex because the literature on collaborative learning displays some confusion. Our concerns focus at the intersection of mathematics education and psychology and it must be admitted that the former community have not yet settled on a well-articulated notion of collaboration, while the latter has largely failed to account for the epistemological dimension, often considering the knowledge domain as incidental. If we are to understand how the presence of others mediates the growth of an individual's web of meanings, we will need to clarify the influence of knowledge, task, setting and medium: it will certainly not suffice to consider 'groups' or 'collaboration' in the abstract. We will therefore begin our discussion by seeking some theoretical clarification of work on collaborative learning, before we return to mathematics.

1.1. *Some theoretical background*

Learning as a result of conflicting feedback from one's actions is a fundamental pillar of Piagetian theory. When an idea contradicts an existing schema, then a state of cognitive conflict exists. Yet as Bryant (1982) points out, this conflict may not necessarily lead to learning, even though it is likely that the student becomes aware that something might need to be learnt or seen in a different way – especially if the conflict is not threatening.

To interpret the potentially facilitating effects of collaborative learning within a Piagetian framework, one needs to extend the notion of conflict to include socio-cognitive conflict: the ability to stand back, *decentrate*, and reflect upon one's own activity in the light of differing points of view. From this perspective, there is a natural tension between the individual and the group (which is assumed to be composed of peers without the presence of any adult authority who might skew the interactions). Piaget stressed the need for differentiation and reciprocity in social interaction, and forwarded the view that logical thought – the result of the co-ordination of differing viewpoints and the reversibility of collective conservation – was very difficult, perhaps impossible to attain on one's own:

> At the stage at which groupings of concrete operations and particularly formal operations are constructed... the problem of the respective roles of social interaction and individual structures in the development of thought arises in all its acuteness.
>
> (Piaget, 1971, p. 162)

Piaget goes on to describe the complementarity of social and individual thought in the quest for equilibrium:

> ...without interchange of thought and co-operation with others the individual would never come to group his operations into a coherent whole: in this sense therefore, operational grouping presupposes social life. But, on the other hand, actual exchanges of thought obey a law of equilibrium which again could only be an operational grouping, since to co-operate is also to co-ordinate operations. The grouping is therefore a form of equilibrium of inter-individual actions as well as of individual actions, and it thus regains its autonomy at the very core of social life.
>
> (*ibid.* p. 164)

The evident importance attached by Piaget to social interaction has often been overlooked. Nevertheless, there has been considerable interest by social psychologists in the study of the development of general logico-mathematical schemas. One finding is that students with differing points of view working together, can be committed to overcoming their contradictory points of view and in the process re-evaluate and restructure their perceptions so as to develop shared solution strategies (see, for example, Doise & Mugny, 1979; Phelps & Damon, 1989). From this perspective it is diversity of view which counts:

it does not matter if 'both partners are equally wrong, as long as they are wrong in different ways' (Light, 1983). However, for individual cognitive gain, the social context of the problem situation has to be structured so that each individual takes account of others' centrations (Perret-Clermont, 1980). This is the first of many indications that the specificities of the activity play a crucial role.

Despite the salience of the conflictual model, it has increasingly been observed that students working together can make progress even in the absence of any apparent conflict, and that conflict is not the only motor of cognitive progress. On the contrary, students often appear to coordinate their mathematical activity interactively by negotiation and argument. For example, Laborde (1994) has shown that in classrooms where groups are formed informally, '...social interactions are less frequently based on well-delineated conflicts ... Proposals made by one student may be improved by the partner and transformed into more sophisticated solutions' (*ibid.* p. 152).

An alternative psychological viewpoint stresses cooperation rather than conflict, emergent consensus rather than disagreement. It resonates with the Vygotskian view which emphasises the interdependence of social and cognitive factors, and recognises a mechanism for cognitive growth as interpsychological regulation in a social world. Vygotsky sees a dialectical relationship between the individual and the group: development emerges from a process of internalisation, the internal reconstruction by the individual of external behaviour through the medium of language. While Vygotsky focused primarily on interaction between a student and a 'more capable' adult, Forman (1985) extended this notion to interaction amongst peers, arguing that where peers negotiate and guide each other through a task, they appear able to develop a shared understanding of the problem.

This perspective seems directly to contradict the Piagetian viewpoint. Yet Kruger (1993) suggests that these apparently disparate theories are in practice not so far apart:

> ... in those studies that claim conflict is essential to cognitive restructuring, conflict is not the simple confrontation of opposing centrations, but an extended discourse that explores the reasoning behind the various viewpoints being presented.
>
> (*ibid.* p. 166)

Kruger points out that researchers who locate their work within a co-operative framework, while they emphasise that agreement is reinforced by clarification and extension, stress the importance of exploring and comparing differing ideas – a scenario not so different from the conflictual situation: 'Whether this interactive situation is described as conflictual or co-operative may be a matter of semantics more than of substance' (p. 167). She concludes:

> Collaborating children are working at the level of ideas; they are finding errors, finding powerful differences, agreeing to disagree, conflicting. They

are also laboring together, communicating their ideas to each other, making discoveries about what works, creating a good solution. Collaborative learning is learning from analysis of the other's perspective, and from the other's analysis of one's own perspective, and from a new synthesis of those analyses. It is both dissection and creation.

(*ibid*. p. 179)

So there is some psychological basis for viewing socio-cognitive conflict and negotiation of joint action as two sides of a coin rather than opposing poles. We might construe groupwork as a domain in which to bring together the individual constructivist and the socio-culturalist; after all, collaborative learners have to see their own views as interweaving with the group problematic, they have to search for a means to reach some sort of negotiated consensus.

There are, of course, formidable theoretical difficulties in any attempt to synthesise the two approaches, not least that posed by Resnick (1991) who wonders: 'How can people know the same thing if they are each constructing their knowledge independently? How can social groups coordinate their actions if each individual is thinking something different?' (*ibid*. p. 2). Fortunately, we do not need to pursue these thorny issues further, for the problems of learning in groups *in general* are of only marginal concern. Our focus remains on mathematical domains, and we now review the state of our current understanding of collaborative learning in mathematical settings.

2. COLLABORATIVE ACTIVITY IN MATHEMATICS

Interest in the potential of groupwork in mathematics stems from many sources. One is the perception of mathematics as a threatening and isolating activity, and the hope that collaboration offers the chance of making mathematics more human. There is a widely-held view that the incorporation of discussion into mathematics classrooms may increase motivation and involvement and, at the same time, induct participants into the formal discourse which many have so much difficulty adopting. Others are still more optimistic: Hatano (1991), for example, suggests that encouraging students to share ideas in collective enquiry is likely to help overcome difficulties arising from the duality of mathematics as both a tool and an object.

These claims have been evident for some time in policy documents such as the Cockcroft Report (Cockcroft, 1982) in the U.K. and the NCTM[1] standards in the U.S.A., which have endeavoured to develop more diversity in teaching style, by including more opportunities for discussion between teacher and students and between students themselves. There is little one could disagree with here, unless one is hopelessly locked into the belief that learning is entirely an individual affair, always best undertaken in isolation from one's peers.[2] But that should not deter us from searching for evidence which takes us beyond the 'groups are good' style of finding. As Good et al. (1992) observe, while there has been an enormous amount of research, there is 'a paucity of process data to indicate what happens during small-group instruction and which dimensions of the model are important ...' (*ibid.* p. 167). Their survey illustrates that although there is ample justification for adopting small groups in mathematics classrooms, it is not necessarily the case that in small groups students become active learners, or that they necessarily increase their verbalisation, critical skills or collaboration. Moreover, teachers use small groups in a variety of ways, one of which, for example, is simply as a management technique to accommodate differences in student achievement rather than to introduce a change in the way mathematics is constructed in the classroom.

It is clear that any reasonable framework for evaluating the contribution of groupwork must take account of at least two broad issues. First, the knowledge domain itself, the specificities of task, as well as the medium in which it is expressed. Second, it is unreasonable to expect groupwork to have any real influence on learning unless the participants have a commitment to sustained argument, justification and explanation: here the role of language surfaces clearly. We consider these two issues briefly, before we elaborate our argument by presenting some of our own findings.

2.1. *Knowledge domain, task and medium*

We have indicated that a useful cut in the psychological research runs between the conflict and co-construction paradigms. Not surprisingly, mathematical studies in this area have tended to inherit their methodology from one or other of these paradigms. So, for example, Bell (1993) studied groupwork as a device to resolve 'conflicting conceptions' and to 'eliminate misconceptions': although the findings were reasonably positive it could not be argued that the addition of groupwork into conflict situations necessarily led to learning. Unsurprisingly, conflict has to be recognised and then reacted to, and these actions in turn will be influenced by the extent of commitment to a particular point of view before any alternative perspective is faced.

Some of our own research indicates the possibility of a perspective which cuts across these coarse-grained findings. We have found it useful to classify disagreements among students into two types: *local* disagreements about what to do and *global* disagreements where at least one student exhibited some sense of a broader underpinning of his/her assertions (the findings which give rise to this distinction are reported in Hoyles, et al., 1991a). It turned out that local disagreements had the effect of destabilising inefficient solution strategies, helping to keep the students on track.[3] Global disagreements occurred rather rarely but when they did, they tended to arise at the point where there was *a need to express ideas symbolically*, a point whose location varied according to whether the medium of expression was computational or pencil and paper. It is global disagreements which simultaneously activate the web and extend the discussion beyond the immediate setting.

This is a telling outcome. It suggests that it is at precisely the point where there *has* to be negotiation, that global disagreements arise. At some point, inter- and intra-peer incompatibilities and inconsistencies in problem structure and solution strategy have to be made explicit. When that happens, global disagreements tend to occur and some kind of co-construction is likely to follow.[4] It is here that mathematical meanings are jointly constructed. It follows that domains of abstraction – in which explicitness forms a natural component – provide a particularly felicitous framework in which conflict *and* co-construction might occur. Conflict arises from abstraction; abstraction develops the need for co-construction.

The creation of a domain of abstraction which encourages expression of differing mathematical perspectives is unlikely to occur without prior planning: and it is here that the 'experimental' school emerges as an important influence. The French didacticians designate by *formulation*, those situations which are inevitably social – situations where mathematics is used and expressed. Laborde (1993b) distinguishes two modalities amongst these: those where the social dimension is *a priori* since an asymmetry between students with respect to some mathematical information is built into the activity and this information has to be 'sent' from one to another; and those where the social aspect is 'added' through the stipulation of presenting a common solution.

Laborde (1994) describes an example of the second modality, which has formed the basis of several experiments. In this approach, students with different and robust conceptualisations of some mathematics are brought together to overcome the contradictions inherent in their viewpoints. In fact, elements of *RatioWorld's* design in Chapter 4 were informed by this and other work of the French school, and we briefly reconsider it here.

First, it is clear that while conflict is a successful mechanism in generating learning in some carefully designed tasks, there are instances where it does not function as planned – such as the reaction to the 'hole in the HOUSE' which simply resulted in patching the hole by edging forward. This illustrates just how difficult it is to engineer situations in which there is a congruence between our aims and those of the students and as a corollary to interpret student behaviour. What makes matters worse is that students display a deeply-rooted tendency to hold contradictory points of view simultaneously; if inconsistencies of approach or strategy are pointed out to them this cuts no ice. They see no reason why operations should not be performed differently on different occasions – we saw this in the tasks of *RatioWorld* in which students who multiplied in proportional questions when the numbers were easy soon reverted to adding as the going got tougher. When the illogicality (from our perspective) of this was pointed out to them they did not share our concern: as we pointed out in Chapter 4, consistency is part of a mathematical culture but not necessarily that of the students.

Even if conflict is recognised, it does not necessarily lead to new knowledge: much is contingent on the task, as well as the nature of the conflict resolution. As Laborde (1994) points out, conflicts are 'not necessarily solved by the construction of a new rule' (p. 151). In fact, some earlier work of our own revealed some children who believed that arguing in school was wrong, and had therefore avoided it at all costs during their groupwork. It is challenging for us now to reflect on just when and where we had expected them spontaneously to distinguish intellectual from personal argument! A similar observation has been made by Balacheff (1986, p. 12) who suggests that when faced with a mathematical problem the students' primary aim is to produce a solution – not to construct knowledge – and this implies slipping away from any cognitive conflict, rather than harnessing it in the service of new learning.

Within the constructivist paradigm of recent mathematics education research, co-construction rather than conflict underpins much of the work. For example, Smith et al. (1993) argue that classroom discussion should attempt to refine rather than to replace, to provoke articulation and reflection rather than confrontation. Other researchers evoke notions of 'shared knowledge' or a 'taken-as-shared' basis for mathematical activity in collaborative settings (Cobb, Yackel, & Wood 1992) in order to avoid any assumption of congruence between different individuals' conceptions. These descriptions reflect the negotiated, non-prescriptive nature of much groupwork research. Yet even here, we find the interweaving of the conflictual with the consensual as summed up by Cobb (1990), describing the role of collaborative learning:

1. Learning should be an interactive as well as a constructive activity –
   that is to say, there should always be ample opportunity for creative
   discussion, in which each learner has a genuine voice;
2. Presentation and discussion of conflicting points of view should be
   encouraged;
3. Reconstruction and verbalisation of mathematical ideas and solutions
   should be commonplace;
4. Students and teachers should learn to distance themselves from ongoing activities in order to understand alternative interpretations or
   solutions;
5. The need to work towards consensus in which various mathematical
   ideas are coordinated is recognised.

(ibid. p. 209-210)

How to operationalise this programme requires careful planning. We know that simply generating conjectures or 'brainstorming' in groups might have little influence on learning. Pirie and Schwarzenberger (1988) report the 'scatter-shot' nature of much that passes in the classroom for discussion, and Stacey (1992) studied a group of children who often generated plenty of ideas but had no mechanism for evaluating them. We are faced with a new version of a recurrent dilemma: connections are made, webs are constructed – but are they the right ones? Do they build a framework on which a mathematical perspective can be built?

To make some progress on this question we clearly have to go beyond viewing mathematical groupwork as a black-box and consider some process characteristics of student collaboration and conflict. Studying the language within group settings provides the most obvious window on these interactions.

2.2. *The role of language in mathematical learning*

Even if we restrict our focus to the language of discussion, there is great diversity in the approaches of different researchers. In order to make sense of this diversity, we start by distinguishing between the 'experimental' and the 'natural' approach to the study of discussion in mathematics. We have already encountered the former, mainly French, approach which involves the evaluation of 'didactical experiments': the mathematics and the group organisation are formally specified and managed (see also Healy, Pozzi, & Hoyles, 1995). In contrast, the natural approach involves the analysis of peer interaction during attempts to coordinate mathematical work as it occurs 'naturally' in classrooms (see, for example, Cobb, et al., 1992; Pirie, 1991). In both scenarios, attention is paid to process aspects: typically, student-student and student-teacher interactions are recorded in an attempt to make inferences with regard to conflict, negotiation and conceptual development.

Language has a rather special role in learning mathematics (this does not imply that other means of communication such as gesture are unimportant) and one approach to thinking about student talk is to take the language itself as a focus, rather than simply thinking of its relation to cognition. Considered in this light, an integral part of the process of identifying mathematical objects and relationships between them, is the development of a language of description: in this case, the challenge for mathematical pedagogy is to aid children to 'speak mathematically' (see Pimm, 1987).

In the field of mathematics education, most research to date has viewed student language as a window on the role of conflict and co-construction in mathematics learning, rather than an object for analysis. There are at least two functions of talk in mathematical activity: the first cognitive (clarifying for oneself), the second communicative (clarifying for someone else). The former enables the speaker to step aside and reflect upon a piece of mathematics; the latter involves explaining and justifying one's strategies and logically rejecting alternative proposals (see Hoyles, 1985). The central tenet is that verbalisation helps students to own their knowledge, to ask realistic questions and to make mathematical structures and relationships explicit. Discussion can, it is argued, assist in focusing awareness between the tool use of mathematics and the appropriation of the relevant structures and relationships. Formulating descriptions of a situation for another forces the speaker to pick out some variables and relationships in any situation at the expense of others. In the effort to communicate, the speaker has to strive to frame his or her thoughts in language which conveys meaning, to try to see another point of view and develop more flexible approaches to strategies and solutions.

It is far from simple to classify verbal exchanges as motivated by the need to communicate or by the desire to represent and develop one's own ideas. More difficult still is the task of mapping the results of this classification on to expressions of conflict or co-construction. Problems are compounded by some confusion as to what exactly constitutes *mathematical* talk. For example, Pirie and Schwarzenberger (1988) define mathematical discussion as having the characteristics of purposeful talk on a mathematical subject in which there are genuine student contributions. But this definition begs as many questions as it answers. How, for example, can the researcher tell whether student contributions are genuine, and whether the argument is purposeful or indeed mathematical?

To make progress, we will need to consider further the relationship between action and language in mathematical settings; this in turn suggests that a computational window will be helpful. There is a burgeoning corpus of literature concerning the co-operative use of computers (see, for example, special editions of *Learning and Instruction* (Mevarech & Light, 1992); and *Social Development* (Howe, 1993)). There are, however, difficulties in integrating the findings from this research, not least because of the wide range of methodologies adopted. Different parameters defining research procedures (for example length of study, nature and timing of tests, age of children, context of learning) go some way to explain conflicting results (see Isroff, 1993).

More fundamentally, the literature displays no uniformity regarding how 'the computer' is conceived, and what is taken as the role of software. In some studies, the computer is simply part of a 'treatment'; software is ignored altogether and its specific characteristics not deemed to have any effect on student interaction (see, for example, Messer, Joiner, Loveridge, Light, & Littleton, 1993). In complete contrast, other investigations use software specifically to structure collaboration – by the manipulation of task presentation under software control, and the restriction of available forms of response (see, for example, Blaye, 1988; Davidov, Rubtsov, & Kritsky, 1989).

We shall propose an alternative to both these approaches, based on our view of the computer's roles which we described in Chapter 3. It hinges around the idea of the software as a mediator of social interaction, not so much as a way of constraining the action, but as a medium through which shared mathematical expression can be constructed and – critically – observed.[5] From this perspective, the nature of the software is not just an important variable; it is crucial, for two reasons. First, it must provide a means by which students can build their own problem space; it must, in other words, lend itself to a constructionist role. Second, it must have the capability to serve a communicative function through which meanings can be made explicit and open to negotiation.

To illustrate our alternative, we now turn to a study undertaken by one of us (Hoyles), Lulu Healy and Stefano Pozzi. In this study, the interaction between individuals became the focus of attention. There is, as we have said, considerable complexity involved in any investigation which attempts to unravel the interaction between setting, medium, and peer interaction. In the study we outline below, we have tried to resist the temptation to control this complexity by the suppression of potentially influential factors, or at the expense of impoverishing the learning situation.

3. A STUDY OF GROUPWORK

The study was part of a three year project, *Groupwork with Computers*, which set out to investigate students working in groups with computers in a variety of mathematical settings.[6] Before the study began, the teachers from participating schools were introduced to the organisation of groupwork in their classrooms, familiarised with Logo,[7] and helped to introduce it into their classrooms. Given the need for this preliminary teacher development phase, the sample of schools and classes could not be random: and in any case, any attempt to 'randomise' the sample would have cut across the grain of naturalistic studies which we favour. Nevertheless, the catchment areas spanned families of diverse social classes and ethnicity which were reflected in the student groups.

A number of group tasks were designed specifically for the purpose of stimulating the learning of mathematical topics which formed part of the school curriculum. In common with all the studies we have outlined so far, our purpose was to study thinking-in-change – specifically, the collaborative process in the course of learning. One objective was to tease out links between task, content, software, group processes and students' mathematical learning,[8] rather than investigating whether groupwork was better or worse than other forms of instruction.

The students who participated in the study consisted of eight groups of six students ranging in age from nine to twelve years. These were distributed among seven classes. We will call this the case-study group, to distinguish it from the 'baseline' group which consisted of all the remaining students in all the classes. Each case-study group consisted of three girls and three boys who came from the same class, so interpersonal relationships had been established prior to the research. The teachers were asked to choose students whom they felt would get on well together: yet in interviews with the teachers, it became apparent that some had difficulty in setting up 'friendly' groups. The groups were selected by the teacher so as to include, in their assessment, a girl and boy from each of a 'high', 'middle' and 'low' attainment in mathematics.

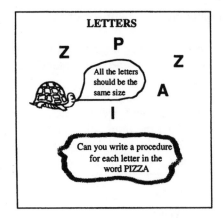 

Figure 6.1. Letters Task.

3.1. *The tasks*

All groups undertook the same three research tasks. Each task comprised a set of activities, called *local targets,* to be shared out by the student group amongst subgroups, and constructed at the computer – three computers were available for the group's use. The students would then bring together the outcomes of this distributed computer-based work in discussion of *global targets.* Our focus here is on two of these tasks, each of which incorporated Logo, but which involved different mathematical aspects. The first – the *Letters task* (Figure 6.1) – focused on the processes of programming; the second – the *Spokes task* (Figure 6.2) – was concerned with ideas of angle. Each task was completed by the eight groups, so data were collected from 16 settings and 48 individual students.

In *Letters*, the local targets comprised building Logo procedures for the individual letters of PIZZA; the global targets were the co-ordination of these letters to produce the word on the screen. In *Spokes*, the local targets were the construction of 'spoke' patterns with different numbers of rays emanating at equal angles from one point, and the global targets involved locating and formalising a relationship between these angles and a complete turn.

At the beginning of each research session, the group of six students was given one copy of the task, each part of which was described in detail by a researcher who also encouraged the students to work collaboratively and produce one group outcome. After this introduction, no further interventions were made. Each session lasted some $2\frac{1}{2}$ hours.

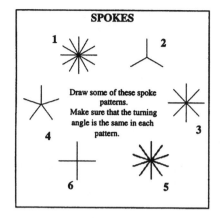
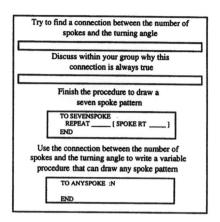

Figure 6.2. Spokes Task.

3.2. *A complementarity between quantitative and qualitative data*

This study marks an extension of the methodologies we described in Chapter 4. Primarily, it elaborates an approach based on our belief in the efficacy of studying complex learning phenomena by a judicious mixture of quantitative and qualitative methods. So in this study, data of two types were collected. First, in order to gain detailed knowledge of student interactions with each other, with the computer, and with the mathematical ideas involved in the tasks, each research session was videotaped. Two researchers were always present, one systematically recording task-based interactions around the local and global targets, while the other made an impressionistic commentary of the group's involvement in the task, the extent of mutual support and exchange etc.

Second, in order to measure any learning associated with the research tasks, a series of written tests was administered individually a week before, immediately after and four weeks after each task. The tests were completed by all the students in the research classes, both case study and baseline. Rather like the written tasks in *RatioWorld* these questions encapsulated what *we* saw as the same mathematics – modularity and rotational symmetry – but posed in a schoolish form. While the case-study students undertook the research tasks, the rest of the class received no equivalent teaching of the mathematics involved nor worked on relevant computer activities; in fact, the teachers were asked not to undertake any class activities on the concepts under investigation during the 6 week period of the study.

Thus, in common with the methodology we employed in *RatioWorld* (Chapter 4), there was an intention for quantitative and qualitative methods to complement each other: for the former to aid in the comparison of the success of similarly structured groups working on the same tasks; and for the latter to develop an explanatory framework which could throw light on the complexities of interactions between students and activity structures. We might note that the dialectical relationship we perceive between quantitative and qualitative data collection,[9] reverses a common claim of those within the research community who employ qualitative techniques for 'hypothesis generation' and quantitative approaches for showing 'effects'.

3.3. *Gross trends*

The initial thrust of the analysis was both qualitative and quantitative; detailed case studies of the different group settings were developed, and multi-level regression[10] was applied to the raw score test data in order to model progress in general and to examine the influence of key background variables (details of these and other aspects of the study, are given in Pozzi, Healy, & Hoyles, 1993, on which the next few paragraphs are based). Multi-level regression was chosen because, unlike comparable statistical methods, it takes into account the inbuilt hierarchy and clustering of the data, in order to estimate the regression parameters: a further illustration of how it is now possible to use statistical techniques in a sensitive way which respects the complexity of the data set. The (two-level) hierarchy in our study comprised individual students, and students within classes. The class level of the model is important, since irrespective of whether a class was case-study or baseline, students within a class shared common experiences which would be likely to result in correlation of their scores (because this effect is difficult to measure, researchers often ignore it altogether). The method takes into account, therefore, variance between classes as well as between students.

First, we checked to see if there were any differences between the pre-test scores of the case-study and baseline students. Although some could be identified, we were able to conclude that the case-study students were representative of the rest of the class in terms of understanding the mathematics of the tasks. Second, it became apparent that there were discernable differences in progress between the case-study and baseline students which could be examined statistically.

For each task, we first assumed that the delayed post-test scores could be modelled as a linear function of pre-test scores and involvement in the groupwork. Then we elaborated each model, to explore the influence on individual progress of gender, pre-test attainment and teacher-designated ability.[11] Applying the basic regression model,[12] our findings suggested that students within the case-study groups improved significantly more than the rest of the class across both tasks – both in post- and delayed post-tests. Table 6.1 shows the estimated parameters of improvement at delayed post-test. Mean benefits due to involvement in each task were: *Letters* (8.2%) and *Spokes* (12%).

Even though the students in the rest of the class were not engaged in any comparative instruction, individualised or otherwise, the considerable

Table 6.1. Regression estimates for basic model. * p < 0.05

Task	Regression estimates and standard errors			Variance	
	Intercept	Pre-test	Groupwork	Between-class	Between-student
Letters	52.3 (2.18)	0.8 (0.063)	8.2 (3.64)*	0	333.88
Spokes	56.0 (2.71)	0.7 (0.057)	12.0 (4.57)*	0	593.88

improvement of the case-study students in comparison with the baseliners is interesting for two reasons. First, the tests were not simply individualised versions of the group tasks, but placed the mathematics in different contexts, both in terms of being exclusively paper-and-pencil based and in the way the mathematics was expressed. Second, the improvement was sustained and in some cases even increased at delayed post-test – in contrast to the more common drop in progress at delayed post-test.

We then applied the different regression models to examine whether there were any associations between progress on each task, and gender, pre-test score and the teacher-designation of high, middle and low ability. These all proved non-significant: there were no discernible gender differences in either pre-test attainment or progress to post and delayed post-test. More importantly, there were no discernible gender differences in progress amongst the case-study students themselves as a result of involvement in the group settings. Similarly there was no evidence to suggest that improvements were related to ability levels as assessed by our two measures.

One aspect did, however, emerge: from the case studies, we found that the effect of antagonism in a group significantly affected the performance of individuals within it. This was only to be expected. But there was a puzzling corollary: the effects of such antagonism varied across tasks. Whereas in *Spokes*, the presence of antagonism within groups was detrimental to learning, the reverse was true in *Letters*: in fact, the inclusion of some mutually hostile members of the group seemed to produce an increase in performance measured by the tests! What was happening?

3.4. What was happening inside the groups?

To answer this question, it was necessary to form a much clearer picture of the processes within groups, a question which necessarily demanded a swing away from the quantitative and towards the qualitative. The aim was to combine all the data to build a multi-faceted description of each setting, from which could be drawn common and contrasting aspects of the group dynamics.

The key methodological tool was based on a classification of all on-task interactions into *episodes* – distinct student interactions with a clearly identified focus which was either a local or global target. Each episode incorporated data on its focus (what part of the task or subtask was it centred on), where it was located (on or off-computer), and what individual students were doing (or not doing). When these episodes were examined in detail, it was possible to discern key process descriptors, which encapsulated how each group managed local and global targets; whether local targets were shared out or replicated by subgroups, and the degree to which global targets were shared. This enabled a realistic comparison of settings to be made.

The pattern which emerged distinguished between two major styles of group organisation. The first, integrated style, involved a sharing of local targets across subgroups, and a high level of global target sharing. In contrast, the second, fragmented style emerged, in which local targets were replicated and there was a low level of global target sharing.[13] The two styles illustrated schematically in Figure 6.3, showed some interesting divergences. In integrated groups, subgroups maintained channels of communication throughout the local and global tasks. At the same time, cross-computer activity – in which students shared what was on each others' screens – was evident. We might say that the web of the support structures was robust, stretched between the computers, and that the whole group contributed to work within a joint problem space. In contrast, fragmented groups became focused on their own work, communicating minimally if at all except through action. As a result, there were low levels of cross-computer activity, and helping across groups with local targets was rare.[14] As we might expect from a situation in which the web of the support system was more fragile, fragmented groups showed a higher susceptibility to diversions and off-task activities.

Our characterisations of organisation have all focused on the group as the unit of analysis. When we turn our attention to the role of individuals within groups, two more elements of the picture become clear: first, the integrated style of working was associated with a democratic pattern of interaction in which the majority of students tended to negotiate their contribution, while a fragmented style was more likely to be dominated by a subset of individuals within subgroups or in the group as a whole; and second, a range of task-related influences began to emerge.

3.5. *Individual learning*

To try to assess more directly the effects of participation in the group tasks on individual learning, we needed to swing the methodology back towards a quantitative measure of individual progress.[15] Here there was confirmation of the initial disaggregated trend: in *Letters*, individual members of the case-study group working in a fragmented group style out-performed baseliners. Yet in *Spokes*, there was a strong association between progress on the tests, and

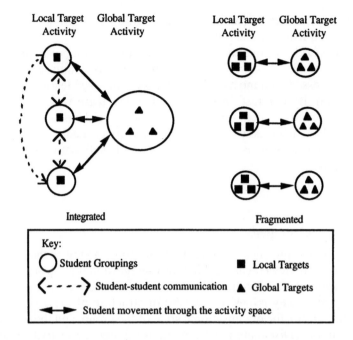

Figure 6.3. Integrated and fragmented styles of group organisation.

working in an integrated style. Why should integration be good for *Letters*, while fragmentation was better for *Spokes*?

To answer this question, we focused exclusively on the case-study groups. Considering *Spokes* alone, it was still true that students working in an integrated style seemed to learn more than those in fragmented settings, but the effect for *Letters* disappeared (although the numbers were small), as did any statistically significant interaction of style with interaction pattern (democratic or dominated).

So we turned once again to an analysis of the qualitative data which revealed something not evident from the quantitative analysis. The quantitative analysis inevitably conflated a range of mathematical ideas contained within the tasks; it is true that these ideas were related to each other in terms of mathematical content, but the total scores on tasks (and groups of tasks) could not shed light onto the nature of learning. For that, we required a much more fine-grained measure of the content of what was learned. So our next step was to construct learning profiles for individual students on the basis of subsets of test items. By analysing these profiles, we were able to characterise each student with respect to a single concept as a *knower* (someone whose scores could not improve as they were high initially), a *non-learner* (scores remained low), a *learner* (scores shifted from low to high) or a *consolidator* (scores shifted from middle to high).

When we excluded the knowers (whose scores could not improve), there was a clear trend. In *Spokes*, over half the students within integrated-style groups progressed, and most of this progress comprised learning rather than consolidation. By contrast, in fragmented styles, there was no evidence of consolidation and a much smaller proportion of students were learners. In *Letters*, students who had worked in integrated styles made considerably less progress, and what progress there was, was evenly split between learners and consolidators.

3.6. *Summing up the study*

We will try to interpret these findings from the point of view of the literature on groupwork we reviewed earlier. First, any suggestion that 'groups are good' is not well-founded. A 'group', like a 'computer', is under-defined. Working in a group hides key issues: the degree of structure of the group activities (high in the study), the organisational style of the group, and the interactional patterns within it. Disentangling the relationships between activity, interaction and type of learning is no easy task and we cannot pretend to have done more than throw some initial light onto the problem. Given some of the extravagant claims made for 'groupwork' and 'discussion', recognition of the complexity itself seems worthwhile. What we have certainly found is that studying groupwork around computers has highlighted, even exaggerated, the diversity of factors that can affect the trajectory of mathematical meaning-making. Ignoring, for the moment, the complexity of the situation, we can conclude that insofar as individual learning occurred, the gains were relatively robust and long term and that structured activities of this kind do not appear to disadvantage particular groups of students: we found no differences across gender or for different abilities. At the very least, the study suggests that in any specification of 'good group settings', the particularities of the learning goal are crucial; we can neither ignore them, nor subsume them under the general heading of 'mathematics' or 'problem-solving'.

The most significant finding is that effective learning of conceptually-based material involving the appropriation of mathematical relationships occurred where there was a synergy of interdependence and autonomy through active construction at the computer which fed into wider group discussion – whether the individual was a vocal or non-vocal participant in that discussion or not.

Equally revealing is the finding based on the *Letters* task, an activity in which the concepts (procedure/subprocedure) were rather directly generated from computer-based actions and reflection upon their meanings. Here, a more fragmented way of working seemed to be quite acceptable – even beneficial. Working closely at the computer, the development of an intense relationship with the medium appears to have afforded just the kind of experience which engendered deep learning.

What are the broader implications for social interaction in mathematical learning? A standard view might interpret the adoption of an integrated style as implying that student goal structures were focused on the sharing of the solution process rather than on individual success. Yet such an explanation does not account for the whole story. In an integrated style, global targets were addressed away from the computer and generally synthesised from a discussion of the computer work derived from different local target activities. Students had the opportunity to reflect on their own constructions alongside the constructions of others. Conflicting strategies were identified and described and the computer used to mediate these descriptions.

This mediating role of the computer provides a window onto collaborative learning in general. It suggests that a purely conflictual or co-constructive model is insufficient to explain student interactions and the development of their mathematical understandings. In our study, it seems that both balanced co-construction with the computer and the co-ordination of others' perspectives in whole group discussion had roles to play in the mathematical learning. It seems that without co-construction, students may not develop any approach to the problem or any language to describe their strategies. Without conflict, students may remain centrated on their own way of understanding the problem.

We hinted at the outset that a candidate for a mechanism which drives learning in collaborative settings might be the process of symbolisation, the construction of symbolic 'code' and the associated domain of abstraction it provides for shared meaning making. And so – given the indications of the *Letters/Spokes* study – it turns out. As students who adopted an integrated style constructed their solutions at the computer, they made sense of the mathematics of the task at their own level of sophistication, and in ways which required them to clarify and formalise their ideas. Therefore, when synthesising local components in the whole group, students had a way to describe their own contribution and were in a better position to take on the mathematical ideas of others. In short, the computer gave them a medium for talking about what they had constructed, comparing their constructions, and refining their conclusions by articulating and reflecting upon their own and each other's strategies. Students working in the integrated style, jointly breathed life into a web of meanings that spanned across their collective and individual constructions.

This balanced scenario stands in contrast to the cursory treatment accorded to their peers' work by students working in the fragmented style. The paucity of cross-computer activity meant that students tended to concentrate on the strategies developed at their own computer and were less likely to be confronted with alternative perspectives. Yet when the outcomes of the task and the conceptual learning associated with it were more technology-based, neither comparison of ideas away from the computer nor mutual decision-making at the computer were crucial for progress: the local webbing of computer interaction sufficed. Provided the task is right, a fragmented style can be effective for individual learning.

4. OPENING NEW WINDOWS

We end this chapter with three general observations, which will point in the direction of the chapters that follow. The first involves methodology. It is clear that the domain of computers, groups and learning mathematics is complex, and attempts to simplify the setting in ways which would lead to simpler methodologies would necessarily lead to an impoverishment of the learning situation. Given this complexity, a reliance on either a qualitative or quantitative approach would probably be overly-restrictive: we have used quantitative data to generate questions and indicate possibilities, and we have employed qualitative techniques to provide explanatory frameworks for what we think we have seen. Although shifting between qualitative and quantitative analysis is not unproblematic, it has allowed us to open windows at different levels of generality while simultaneously clarifying the assumptions we have made at each level – which data, and what kind of data were being utilised or ignored at each phase, and what methods were appropriate? These assumptions are often left implicit, especially in paradigms that are well-worn to the point of becoming clichés.

Our second observation underscores how the material of this chapter fits into the framework we have constructed. We have focused our attention on computational environments, and we have shown how the expression of ideas within the medium could act to extend an individual's web of meanings, either by connecting the individual to quasi-mathematical abstractions built on the screen, or by or catalysing new knowledge-connections between individuals. Equally, we have shown how the ways in which students responded to each other and the computer's structuring of that response, were mediated by the specificities of the task at hand. The pedagogic implication is clear: if we wish to design tasks which exploit the potential of co-constructive collaboration, we had better pay careful attention to the epistemological basis of our activity structures.

Our third observation concerns the role we see for the computer in what follows. To what extent was the computer a central element in our analysis? Put bluntly, are our findings applicable only to computationally-mediated environments?

The answer is in two parts. First, we are certain that the specificities of the computational medium play a critical role in the co-construction of meaning. diSessa (1993) goes so far as to suggest that certain kinds of software (i.e., *Boxer*) afford a 'collaborative spirit'. He lists four ways in which this is evoked: visibility, students can see all aspects of the system; *pokability*, programs can be taken apart and tried out in pieces by pointing to a line and executing it; *structure* which assists in the organisation of the code; and *visible manipulable units*, a program can be pushed about as a unit or split up into subunits any of which can very simply 'travel' as tools in different projects.

We agree with diSessa's general point: the medium may not be the message, but it certainly affords access to certain kinds of messages (and correspondingly, closes down others). In this respect, it is not only the existence of the computational medium that provides expressive power to the learner, but the particularities of how that expressive power is operationalised in the system. Others, such as Scardamelia and Bereiter (1991), have gone further and designed computer learning environments specifically to support co-operative commenting in the context of knowledge-building activities.

There is a second part to our answer. Our discussion in this chapter has relied on the computer to highlight the role of expression, symbolisation, rigour in the (co-) construction of mathematical meanings. As a thought experiment, we could imagine removing the computer from the picture. To be sure, this would entail new analyses, new delineation of the ways in which webbing may or may not occur. If we change the structures of the intellectual and physical resources available, it will not be surprising if the specificities of student-medium and student-student interactions change in subtle and not-so-subtle ways.

But there is another side to this thought experiment. We have seen that the computational medium highlights the process of meaning-making in general (it illustrated, for example, how intimately learning was related to task, and how sensitive it was to the activity structures).

Can the computer open new windows onto more general questions of mathematical meaning-making? Can the computer, for example, help us to see beyond the study of single classrooms, to a more general, systemic appreciation of the ways mathematical learning is constructed in the broader educational culture, at the way in which the educational system comes to construct meanings for mathematics? If we want to further explicate the webs of meanings which are constructed in learning environments, we will need to take account of the ways that teachers perceive their role in relation to their students, and in relation to mathematical knowledge. Can the computer provide us with a mechanism to investigate this relationship, can it afford the opportunity to understand how teachers come to construct meanings for the mathematical knowledge they teach? Finally, the elaboration between cognitive and socio-cultural facets of mathematical learning will involve looking beyond the individual learner, his or her collaborators, or the classroom in which learning and teaching takes place. Can we employ a computational window to gain a view on the ways in which schools – rather than individuals – make meanings for mathematical learning?

Our answer to these three questions is in the affirmative, and the following three chapters provide justification for our assertion. Our focus will centre on Logo, if only because this represents the most elaborated corpus of literature on the relationship between computational media and mathematics in educational settings. In Chapter 7, we will look at the ways in which some corners of the educational system constructed meanings for Logo, and with it, the mathematical ideas it was intended to illuminate for learners. In Chapter 8, we will try to understand what meanings teachers created for the computer in their classrooms (mainly, though not exclusively, Logo programming) and how they came to view the relationships between the computational media and the mathematical knowledge they were responsible for teaching. Finally, in Chapter 9, we will synthesise our new-found insights derived from the preceding chapters, to tell the complex but illuminating story of the introduction of Logo into a school.

NOTES

1. National Council for Teachers of Mathematics.
2. It is interesting to ponder how the newly-fashionable (and cheap) penchant for 'distance learning' interacts with the equally fashionable trend towards collaboration.
3. Light and Blaye reached a similar conclusion about the role of disagreements when students were working interactively with a computer adventure game (Light & Blaye, 1990).
4. There are, of course, other alternatives: e.g., no solution is posed, or one or more participants are ignored.
5. A similar perspective has been proposed by others; see, for example, Roschelle (1992)
6. Details can be found in, for example, Healy, et al., (1995); Hoyles, Healy, & Pozzi, (1992). This research was funded by the Economic and Social Research Council, Grant No. 203252006.
7. Other software was also used, but we will concentrate exclusively on the Logo work.
8. We also examined of the success of settings in terms of group productivity, based on what the groups produced together (see Hoyles, Healy, & Pozzi, 1994).
9. Note that this refers to data *collection*. We try to avoid any *a priori* stance towards numerical *analysis* of data, which, in our view, might be equally appropriate to data gathered qualitatively *and* quantitatively (and equally inappropriate).
10. The statistical package used was ML3, developed as part of the Multilevel Modelling Project at the Institute of Education, University of London (Goldstein, 1987).
11. This was done by adding terms into the model for the following variables: gender; high, middle and low teacher-designation of ability; interaction of gender and groupwork; and interaction of pre-test score and groupwork.
12. The basic regression model adopted was $Y_{ij} = a + bX_{ij} + cT_{ij} + u_j + e_{ij}$ where Y_{ij} was delayed post-test score of the ith student in the jth class, X_{ij} was the pre-test

score of the ith student in the jth class, T_{ij} = 1 or 0 depending on whether or not the ith student in the jth class was involved in the groupwork. The residuals were modelled by two deviation variables; u_j was the deviation of the jth class and e_{ij} was the deviation of the ith student in the jth class. This allowed between-class and between-student variation to be shown separately. In Table 6.1, therefore, between-class effects were modelled as u_j, and between-student effects as e_{ij}.

13. 10 out of 16 settings were integrated. 5 out of 16 were fragmented. In fact there was a third style (connected) which involved a reasonable level of global target sharing, but local targets were replicated rather than shared. There was only one such setting, and we ignore it here.

14. Whenever a fragmented style was adopted, exclusively single-sex groups were formed – a marked contrast to an integrated style.

15. Once again, we used multi-level regression analysis.

CHAPTER 7

CULTURES AND CHANGE

National Curriculum Attainment Target 3, Level 2:
Explore number patterns.
National Curriculum Attainment Target 3, Level 6:
Explore number patterns using computer facilities or otherwise.
(Department of Education and Science, 1991).

1. INNOVATION AND INERTIA

There is a substantial literature which has explored educational innovation and change from political and cultural perspectives (see, for example, House, 1974; Rudduck, 1991). It indicates clearly that even when innovations are generally considered to be desirable, it does not follow that they will be successfully 'implemented' in the classroom – as Fullan announced bluntly in his AERA address in 1989: 'All reform efforts to date have failed' (Fullan, 1989). It has become commonplace to see educational systems characterised by inertia, difficult if not impossible to alter speed or direction: 'the more things change the more things stay the same' (Sarason, 1982, p. 58).

Yet, a few decades ago, it was generally accepted that a combination of good ideas, money and energy from external agencies could quickly and easily transform schools and curriculum. One example was the introduction of the computer-assisted teaching system, PLATO, into some community colleges in the U.S.A. In a fascinating case study, House (1974) traces the gradual disintegration of this innovation under the combined influences of a multitude of factors: lack of clarity of the change process, naivety in thinking about the translation of objectives into practice, internal politics and conflicts between groups, technical problems, lack of resources and limited teacher preparation. This was one of many spectacular failures at that time – all well-resourced and arising from sound educational ideas.

The post-war period has been replete with attempts to reform children's mathematical learning. Interestingly enough, many if not all of the initiatives were catalysed by a similar set of slogans: the curriculum is out of date; it does not reflect new ideas about mathematics and about the way it is learned; it does not help children develop to their full mathematical potential; it produces negative attitudes to the subject amongst certain groups (such as girls and minorities); and finally, because of the importance of mathematics in today's society, the poor competence in mathematics amongst the population at large has far-reaching consequences for national wealth and competitiveness. A 'solution' is then proposed; these range from the most ambitious, the creation of a new curriculum (e.g. 'new maths'), to the more modest introduction of new topics (e.g., graphs or logic), or the advocacy of new approaches (problem solving, or more recently, using 'the computer'). Whatever the solution, its standard trajectory begins with panacea and ends with disappointment.

We will have much to say about this process, not least the standard constructs which are often used to 'explain' this pattern. Before we begin our task, we give an illustration of the process we hope to describe, by briefly charting the passage from birth to adoption of one initiative in U.K. school mathematics.

Following the Cockcroft report, *Mathematics Counts* (Cockcroft, 1982) with its recommendation that mathematics teaching 'should include opportunities for... investigational work' (*ibid.* para. 243), there was strong pressure to introduce investigative learning into mathematics classrooms supported by central government, local education authorities (LEAs) and by the majority of teachers' organisations. The main vehicle by which this change was to be introduced was through extended coursework in GCSE mathematics[1] for 16-year-old students, which was initially to include two types of activity: the exploration of 'abstract' mathematical situations and the tackling of 'real' tasks in which mathematics was embedded in activities from the world outside school.

The goals of the innovation appeared to be clear enough; to stimulate a new way of working in the mathematics classroom which would give students a sense of purpose and foster their creativity and autonomy. Yet in the move into the classroom the planned changes became transformed. Investigative work became separated from mainstream mathematics; it became circumscribed within a new topic, the 'investigation', and this began to take on a certain uniformity. 'Real world' tasks became less and less real, and an investigation came to be defined as a type of project with a remarkably common structure – collect some data from examples, find a pattern, make a generalisation and extend it, if possible, beyond the original problem situation. In parallel with the development of this routine, a set of 'rules of the game' emerged which laid out what had to be done to obtain a good mark – write a narrative of plans thoughts and actions, draw up a table, do some algebra if you possibly can. So, largely as a consequence of the requirements of assessment, investigations became institutionalised and lost their investigative character. The ubiquitous data-pattern-generalisation investigation which reduced all mathematisable situations to numerical models and sought no justification beyond numerical argument, became drills of a new kind – in direct opposition to the original intentions of the reform and in the process lost their mathematical integrity. As Morgan (1995) has argued: 'the ideals of openness and creativity, once operationalised through the provision of examples, advice and assessment schemes, become predictable and even develop into prescribed ways of posing questions or 'extending' problems and rigid algorithms for 'doing investigations' (*ibid.* p. 408).

The main problem with this and similar 'reforms' in school mathematics is the simplistic formulation of the 'problem' and the 'solution'. In the 'delivery of the curriculum' approach, so pervasive in school mathematics initiatives, it is assumed that children and teachers, the school culture and even the population at large will not interfere with the goals of the innovation. The 'change agents' appear not to recognise or will not admit, despite all the evidence to the contrary, that changing the mathematics curriculum will inevitably have unforeseen interactions with classrooms and institutions.

The simplistic explanation is that 'the system', 'the school' or 'the teachers' *distort* the innovation. In its implementation, the argument runs, those who do not understand its purpose or share its aims wrench control of the innovation from the innovators, and in the process, warp its intentions and objectives. But schools and classrooms are not inconvenient obstacles; they are settings in which cultures are continually renovated. Change involves change in cultures, and innovations which see themselves standing above cultures are doomed to failure.

From our point of view, understanding the process involves studying how individuals, groups and cultures co-construct meanings in the course of change. In this chapter, our focus is on knowledge; we will look in two directions through the window of the computer – specifically the Logo computer – at the ways in which the epistemologies of Logo and LogoMathematics shaped and were shaped by the educational system and by the research community.

Our initial focus on epistemology has an important corollary. It allows us to believe that we can contribute to the study of curriculum change in an academic field already crowded with sociologists and scholars of curriculum theory. There are two reasons for our optimism. First, we regard the content domain, mathematics, as a fundamental component of the change process; if there is a dialectical interaction between the epistemological and the social, between the ideas underlying an innovation and the setting into which it is introduced, then the specificities of the knowledge domain must be important to our understanding of the change process. Second, our experience is from the inside – we were part of the Logo story, and are reasonably well-positioned to trace the subtle shifts in emphasis and focus during Logo's move into the practical arena. Our hope is that the ways in which Logo was moulded and transformed by the system provides us with some insights into the meanings of the change process for school mathematics more generally.

2. VISIONS OF LOGO

Conventional wisdom asserts that the computer has not achieved the radical effects that its proponents believed it would some ten or twenty years ago. As Becker (1982) has put it:

> There were 'dreams' about computer using students... dreams of voice-communicating, intelligent human tutors, dreams of realistic scientific simulations, dreams of young adolescent problem solvers adept at general-purpose programming languages – but alongside these dreams was the truth that computers played a minimal role in real schools...
>
> (*ibid*. p. 6)

Now in the nineteen-nineties, it is important to understand how much of that early truth still remains. At the same time, we need to understand to what extent the idea of transforming teaching and learning through computers remains plausible, and how little time there has been to develop the tools – both technical and pedagogical – which might bring about this kind of radical change.

In mathematics, over the past decade or so, every genuinely innovatory piece of software has been greeted with a mixture of excitement and apprehension. Amongst the enthusiasts, there is joy at the possibility of a new approach to mathematics, the chance to challenge the straightjacket of the curriculum. On the other hand, this is matched by concern for standards as moves are made into uncharted territory. A decade later, very little has changed – after the surge of excitement, the innovation has either been dropped or smothered as the system wobbles back to its previous equilibrium state. Educational discussions, which in the early phases revolve around the advent of new mathematics or new mathematical practices, become centred on 'overhead' and 'curriculum fit'. As we shall see, the case of Logo provides some insight into how this has happened.

2.1. *Snapshots of Logo*[2]

We begin with some general observations which characterise the ebb and flow of the inertia cycle. The story is complex, not least because it varies across countries; although in most (but not all) developed countries the wave of initial enthusiasm for Logo is over. For example, in France, Logo has essentially died out, despite the heady days of the early nineteen-eighties in which Logo was seen as an educational and political salvation, and Papert and his colleagues ran the *Centre Mondiale* in Paris. To talk of Logo now in France is as unfashionable as it was once *de rigeur*.

In Germany, in a fascinating twist of the inertia cycle, all the chapters of the story were written at the level of theory, with little if anything happening in terms of classroom activity. Logo was hailed in the pages of books and journals: and rejected in the same forum. When *Mindstorms* was published in Germany in 1982, it was hailed by Otte in his lengthy foreword as one of the best books on 'mathematics didactics'. A long and thought-provoking review of *Mindstorms* written by Hans Niele Jahnke from the University of Bielefeld appeared in the international journal *Educational Studies in Mathematics* in 1983 (Jahnke, 1983). This article managed to address the heart of Papert's agenda. It concluded:

> This book should be read by everybody concerned with teaching mathematics. It is exceptional because of its broad views. The efforts at reforming the mathematics curriculum in the last 20 years have shown that no progress can be achieved without such general perspectives. Beyond that, personal involvement makes this book a reading matter which is very inspiring and provokes the reader to reconsider his own fundamental views.

> (Jahnke, 1983, p. 100)

Other books and research papers followed, sowing seeds for academic debate and discussion. This finally reached the pages of the *Journal für Mathematik-Didaktik* and culminated in a hundred-page critical article – perhaps disparaging is a better word – by Bender (1987). In it, Bender launched a vituperative critique of Papert, the Logo 'philosophy' and the educational damage that would follow from Logo's adoption. The paper was noteworthy for its length, style and tone which included polemic and irony not usually found in academic papers, least of all in Germany.

Reactions to Bender were immediate in the form of two 'replies' in the next issue of the journal; but Bender's paper proved to be the more effective and became a standard reference for those who followed Bender's lead in later debates. Shortly afterwards, there came a deafening silence in Germany concerning the question of Logo. Its fate was sealed: those who had previously been its advocates either maintained their silence or were easily ignored through being labelled as 'Logo people' – a term of abuse meaning irrational and biased.[3]

In the U.S.A., Logo moved from being 'in fashion' (for almost everything, including mathematics, problem solving skills, and cognitive 'enhancement') to 'out of fashion' (for everything). After tremendous acclaim in the early eighties, with Logo being advocated and indeed used throughout the curriculum, there has been a relentless backlash. Little is now written about Logo; indeed it seems its possibilities are not appreciated,[4] and teachers and developers can sometimes be seen apologising for or even hiding their interest in it. One consequence of this pendulum swing has been the isolation of 'Logo ideas'. They are interpreted as simply of relevance to 'Logo people' and without interest for the wider educational community – even the mathematics education community. Evidence for this assertion can be gleaned by simply browsing through mathematics education journals and noting the lack of references to work with Logo in the context of broader mathematical themes.

But even in the U.S.A., Logo has not disappeared. It has been reborn as *Microworlds Project Builder* with a new interface incorporating many aspects of direct manipulation – turtles can be dragged to change their position and heading; there is an integrated paint program, and animation is top of the agenda. The change of name is clearly significant – Logo was no longer a marketable commodity in the nineteen-nineties. And the software, whilst being Logo-like, has clearly targeted the non-mathematical audience aiming to exploit a culture of drawing and games: quite complex projects of this kind can be undertaken with little programming. Logo, whose roots are as a programming language to construct conceptual frameworks for mathematics, has been recontextualised as an attractive project-building tool; the programming power, much of which remains, is hidden behind an 'acceptable' face. What, we wonder, determines the limits of acceptability?

2.2. *Logo in the U.K.*

The U.K. case shares many of the elements we have seen elsewhere; yet there are interesting differences, and we will spend a little time looking at these in some detail. In the late seventies, the programming language BASIC was popular. There were claims for the importance for learning mathematics through writing algorithms to make procedures and structures clear and explicit. By the mid-eighties, the rhetoric had changed with the introduction of the notion of 'mathematical programming' – a compromise formula to allow discussion of Logo, a new and apparently more radical alternative to BASIC, without actually having to name the language! Eventually Logo came into its own, quickly followed by spreadsheets, then databases. Now, in the nineties, dynamic geometry software and computer algebra systems are fashionable. Yet Logo survives in two forms: as an elementary drawing program in primary schools, and as a medium for mathematical exploration in some secondary schools.

To understand how this has come about, we begin with a little history. When Logo arrived on the educational scene at the beginning of the nineteen-eighties, there was a surge of interest which, although more measured than that in the U.S.A., gave rise to substantial conferences organised to provide a forum for researchers and teachers to meet and discuss the implications of this new software for curriculum and policy. There was enthusiastic curriculum development together with a burgeoning of research projects. Excitement spread throughout the community, although it must be said this was matched by cynicism and opposition from two sources; from those who still advocated BASIC, and those who wanted schools to remain immune from computer use altogether. Provision of computers in schools was entrusted to the Microelectronics Education Programme (MEP), a government agency; in common with many countries at the time, the U.K. government saw their role as equipping schools with machines first, and only secondarily to aid in the process of deciding what to do with them. On the hardware front schools were exhorted to 'buy British', and substantial subsidies were handed out to, in particular, the 'BBC' computer. As a result, there was little incentive for the company who manufactured it to develop a viable Logo – after all, it had invested heavily in its own 'improved' variety of BASIC.[5]

This mildly interesting accident of marketing and economics had some surprising outcomes. It created a serious gap between the sudden flash of interest in Logo's potential, and the ability of children in schools to actually use it. Into this gap stepped a number of 'turtle drivers': simple programs (usually written in BASIC) designed to draw graphics using a screen turtle: the most successful of these was DART.[6] All of these programs allowed the child to drive a turtle using FORWARD and RIGHT, but none had recursion, list processing, proper control structures, arithmetic operations or serious screen editors. Yet some (not, thankfully, DART), happily packaged themselves with the title 'Logo'. Mike Doyle, the current chair of the British Logo Users Group puts it thus:

> DART, not Logo, was the formative experience for teachers in the U.K. – the philosophy without the language through which it was expressed. Consequently, the notion of Computer Language degenerated to 'precise sequences of instructions'. Logo became little more than 'Turtle Talk'.
>
> (Doyle, 1993, p. 22)

In one form or another, 'Logo' was rapidly taken up in the U.K. As early as 1984 the MEP commissioned a report on classroom experiences by an experienced primary specialist and computer 'non-expert' (Anderson, 1986). Anderson's report showed that 'programmable toys, such as Milton Bradley's 'Big Trak', were not distinguished from Logo-turtles; turtle graphics programs such as DART were not distinguished from Logo; and Logo itself was viewed as difficult, expensive, and (possibly) not necessary for doing Logo' (Doyle, 1993, p. 24).

It would be simplistic to argue that it was merely an accident of software availability that led to the Logo programming language being reduced to turtle graphics – with little emphasis on any aspect of mathematics or even geometry, let alone on programming as a means of mathematical expression. It is more a question of teasing out the factors by which an innovation like Logo changes so that it becomes deemed as acceptable to teachers and to the system. Which aspects take hold and which wither away?

In this case two contradictory processes were at work. On the one hand, the child-centred approach which had come to characterise English primary schools resonated with cut down 'Logo': teachers, parents and head teachers could view 'Turtling' as happily fitting into the wide variety of 'child centred' activities which could be found in many primary classrooms. On the other hand, the very success of Logo's assimilation led to its being viewed as 'an activity' in its own right – not a way of expressing mathematical ideas, but a way of operationalising existing priorities by an 'added on' school topic rather than one integrated into the educational setting.

There was a further agenda at work, stemming from the position that 'teachers get in the way of an innovation'. The accepted wisdom of the time was (and continues to be) that primary teachers cannot cope very well with mathematics, so that insofar as they were asked to teach it, the mathematics had to be hidden. Turtling answered the latter agenda rather beautifully – mathematics without any mathematics! Thus turtling became a legitimate part of primary school practice largely by blurring its goals for learning – school inspectors could approve (turtles had been pronounced mathematical), and teachers could relax ('discovery' was a sure road to learning). Doyle reports that in a survey of the many books for primary teachers which emerged at the time, he was unable to find a single one which introduced a Logo operation, or even a reference to Abelson and diSessa's *Turtle Geometry* which was published the same year as *Mindstorms*.

So Logo became a way of ordering turtles around the screen. Turtle-drivers such as DART shaped the attitudes of a generation of primary teachers, and drawing pictures with a turtle became a new curricular topic. Logo became marginalised by its incorporation – everything had changed but nothing had changed.

Some ten years later, we are left with traces of this early history. For example, in the 1991 'National Curriculum', Logo appeared rather strangely in two 'Statements of Attainment':[7]

Target	Statement of Attainment	Example
Using and applying maths *Level 4*	Identify and obtain information necessary to solve problems.	When trying to draw repeating patterns of different sizes using LOGO, realise the need for a procedure to incorporate a variable, and request and interpret instructions for doing this.
Algebra *Level 3*	Use inverse operation in a simple context	Use doubling and halving, adding and subtracting, and FORWARD and BACKWARD (in LOGO) etc., as inverse operations.

From *Mathematics in the National Curriculum,* (Department of Education and Science, 1991)

We need not dwell on the text of these statements. Nonetheless, it is worth noting the nice irony that there is no mention here of Logo in a geometrical context. Why this should be is unclear, although it is interesting that an earlier version of the Mathematics National Curriculum contained several mentions of Logo, including geometrical ones. We cannot of course know what prompted the change, although we find Doyle's version of events at least plausible. Doyle argues that when it was discovered that a private company had a monopoly on Logo for the BBC computer (we recall that this was the computer which schools were encouraged to purchase) all occurrences of the word were replaced by 'Turtle Graphics' or a similar phrase. We could not wish for a better example of the way in which technologies are shaped by social, political and economic forces.

The revised National Curriculum for mathematics in England and Wales (Department for Education, 1995), is generally less prescriptive than earlier versions. There are, however, many exhortations to use the computer (and calculator) under headings such as: 'pupils should be given opportunities to...' rather than 'pupils should be taught...'. Logo is not mentioned by name but careful reading reveals its traces. In 'Shape, Space and Measures',[8] for example, we read that: 'pupils should be given opportunities to use IT devices, e.g., *programmable toys, turtle graphics packages*' (Key Stage 1; 6–7 years; p. 5). This is taken up later: 'pupils should be taught to understand angle as a measure of turn and recognise quarter-turns and half-turns, e.g. giving instructions for *rotating a programmable toy...*'. Computer use for older children involves creating and transforming shapes, working with 'realistic' data and exploring number. The only computer application mentioned is spreadsheets. The message is that programming is 'out' in any form except for infants!

Logo has been marginalised: it might be seen as a diverting occupation for very young children, or neutralised to 'realising the need' for variables, and the use of inverse operations 'in a simple context'. It survives stripped of programming within carefully designed sequences of exercises aimed at well-defined sets of skills. Logo has suffered the fate of incorporation into a canonical curriculum, a transformation: 'in which what was to have been an instrument for exploration is ossified into an object of teaching' (Kilpatrick & Davis, 1993, p. 208).

Yet Logo is still in use in secondary schools; and where it is used, it is overwhelmingly within mathematics departments (at a forthcoming conference of secondary U.K. mathematics teachers, there are no less than three workshops focused on Logo in one form or another). Interestingly enough, one consequence of this transformation of Logo has been that some mathematics teachers regard its use in their classrooms as unproblematic – it is seen as an everyday part of classroom life in much the same way as a calculator or a piece of chalk. This is illustrated in the following extracts from recent interviews with some mathematics teachers in a London secondary school talking about Logo and justifying its use in their classrooms:[9]

I mean I really like Logo and I think its very good for getting them to think about angles and I mean writing programs... I mean logically structuring...

A lot of the reason for using Logo is because it's a very open environment to use maths in and it develops a lot of logical skills and... reflective skills; I think reflecting on where you've gone wrong... students are much more likely to go back and correct their own mistakes rather than depending on somebody else to correct their mistake... I mean that Logo is very different to the others because it's very, very powerful and its very powerful for doing something that is your own idea as well as for reproducing something else...

Alongside this construction of Logo as powerful, logical, and encouraging of autonomy, there is a parallel set of meanings which has evolved: that Logo is a world into which students can enter without too much intervention:

With Logo you can in the first session just start them off and see what happens... As a starter I find Logo quite nice...

So with familiarity, Logo has become constrained, constructed by some as *easy*.[10] The basic idea is that Logo, through drawing with turtles, connects with what children want to do: hence it is easy to use 'as a starter'. It is these meanings that have often gained priority in the school system; the potential to exploit alternative webs of meanings which would support exploration in different mathematical domains has been largely lost: and the evolutionary process has allowed new meanings to survive at the expense of others. Logo is for 'making shapes', and its role as an expressive medium for exploring mathematics has been suppressed. Now teachers may legitimately enquire what role the computer is supposed to play in the process, and why should they bother with all the difficulty and expense? At this point, it might be easier to dispense with Logo altogether, or use it as a neutered shadow of its former self, as a toy for investigating patterns, or a diverting activity for a Friday afternoon.

It is worth reminding ourselves just how far a vision can be changed as it travels through the culture. Here is Sir Wilfred Cockcroft (1994) commenting on the role of programming in the mathematics curriculum, some ten years after the publication of his influential *Mathematics Counts*:

The writing of simple programs should also be encouraged, *as a refinement of previous work on flow charts* (italics added); for most students competence at the level required when using Logo or BASIC, and involving about fifteen operations, should be looked for.

(*ibid. p.* 46)

And, if reducing programming to a refinement of flow charting is not enough, he continues:

In general, the computer should be seen as an additional means of imparting information which can be used at any stage in support of teacher exposition and in class discussions, whether in problem analysis and solution, or in association with practical work, or in investigatory work.

(*ibid. p.* 47)

Logo has been fragmented to fit different niches in the curriculum – it is either content (draw repeated patterns) or process (part of an investigation);

its teaching separated from learning, the critical connection between the two poles disrupted. At every level of implementation, from the first discussions amongst the computer scientists to policy makers and curriculum designers, we have a tension between why the language was designed, how it was viewed by its creators, and what was required in the educational world.

It is not enough to consider Logo, new approaches to mathematical learning, or indeed any innovation as givens, trying valiantly to enter the portals of education, and being rebuffed by old habits and conventions. The process is much more complex: the innovation itself changes, and entry is sometimes gained at the price of relinquishing the original goals for learning. It is not simply that the educational system mobilises to preserve itself from change (although Papert's (1993) image of an immune response system rejecting a virus is certainly attractive). We shall look at this process more carefully in Chapter 9.

In the evolution of any curriculum change, a struggle for meanings takes place, a struggle which is played out in classrooms, schools, educational systems, research communities and political arenas. We have seen how the relative weight attached to each of these struggles can vary, and not always in predictable ways. The German situation was instructive: here the debate was carried out largely within the academic community. Unfortunately, our ignorance of German precludes our commenting further on the ways this debate evolved. But the German experience provides us with pointers to a surprising but highly parallel development in the U.S.A. and more broadly. Here too, we will see how the terms of debate in the research community were influenced by broader cultural presuppositions, and how, reciprocally, these reshaped the meanings of the LogoMathematical idea for a decade and beyond.

3. MYTHS AND METHODOLOGIES: A CASE STUDY OF LOGO RESEARCH

The research literature on programming as a mathematical activity stretches back some thirty or so years. Unsurprisingly, the early literature argued the case for the languages current at the time, and, as soon as BASIC became available (interestingly, around the same time that Logo was developed) there was a range of studies which sought to illustrate its potential as a medium for learning mathematics. We referred to some of this early work in Chapter 3, although a full review would take us too far from our central concerns.[11]

These studies were supplemented in the early nineteen-eighties by a mushrooming of Logo-based research, when Logo became widely available on small computers for the first time; many (but not all) of these turned away from looking at specific mathematical outcomes towards a more general, problem-solving, orientation. It is, perhaps, somewhat unfortunate that the wide availability of Logo in the early eighties coincided with a massive wave

of enthusiasm for 'problem solving' as the latest panacea for mathematical learning. It was unfortunate because it gave vent to a surprisingly large
number of studies which were inconclusive and/or methodologically suspect.

The most revealing, and certainly the most quoted, research on the 'effectiveness' of Logo in practice was undertaken by a group centred around Roy
Pea and Midian Kurland, who wrote a series of papers which were (and continue to be) interpreted as proof that 'Logo did not deliver what it promised'.
It is time to revisit Pea and Kurland's studies, in order to open a window on
their methodologies and findings, and to appreciate how the mythologies of
Logo research came to be created.

3.1. *Pea and Kurland's approach*

It may seem strange that we have waited some ten years before critiquing the
work of Pea and Kurland in any detail. We offer no excuse for this delay;
moreover we recognise that it is easy to be critical with the benefit of hindsight.
Our rationale for this belated critique is threefold. First, it is never too late to
offer a serious critique of an influential study; and, after the studies we report
in Chapters 4 and 6, we can hardly stand accused of an ideological objection
to quantitative research methodologies *per se*. Second, research of the kind
undertaken by Pea and Kurland continues to inform the scientific literature
in terms of the kind of experiment undertaken (see, for example, Simmons
& Cope, 1993). Third, and perhaps most importantly, Pea and Kurland's
work continues to be cited as an authoritative source. Authoritative sources
deserve serious critique and reflection, especially when it aims to throw light
on broader questions of change in educational cultures.

Pea and his colleagues published many papers on the same theme. Our
comments are drawn from a careful review of much of the published work of
Pea and Kurland, and particularly three papers whose titles we will abbreviate
to PS, TS, and CE, as follows:

Logo Programming and the Development of Planning Skills (Pea & Kurland, 1984a) (PS)

Logo and the Development of Thinking Skills. (Pea, Kurland, & Hawkins,
1987) (first published 1985) (TS)

On the Cognitive Effects of Learning Computer Programming. Reprinted
as Chapter 8 of Mirrors of Minds: (Pea & Kurland, 1984b). Page numbers
refer to the reprint. (CE)

The overarching goal of the authors was to explore the 'effects' of learning
Logo programming. The papers which give the best idea of how Pea and
Kurland set about this are TS and PS. As one would expect, the researchers
held certain prior assumptions. These were: that programming necessarily

involves pre-planning; that if such planning takes place, it ought to 'transfer' to other contexts; and that learning Logo involved a pedagogy which involved little or no intervention.

There are two explicit formulations used by the researchers which allow us insight into the motivations for their research. First, in PS, Pea and Kurland stated in the first paragraph:

> It is well recognised that transfer particularly of intelligent performance to novel task environments is the hallmark of human cognition. To date no studies have adequately demonstrated this transfer of mental rigour however defined.
>
> (PS, p. 1)

Thus the researchers acknowledged that they were looking for something which had not yet been located, and which they were sceptical existed. It is, of course, easier to share their scepticism some ten years later, when beliefs about transfer have been widely called into question by the psychological community. Second, the researchers clearly saw their role as challenging the Logo 'bandwagon', the perpetuation of myths. For example, in PS, they referred to 'the current wave of opinion' supporting programming in schools and referred to schools placing 'such a heavy emphasis on programming because of its *presumed* impact on thinking skills beyond programming activities' (our emphasis). Thus it is apparent that this research study had a clear political agenda. The influences which gave rise to it, namely the apparently overblown claims for Logo, need to be appreciated in the evaluation of the study.

So far so good. The researchers wanted to hold up to scrutiny some current wisdoms concerning education. But the wish for balanced research rather quickly gave way to an altogether more polemical tone when they discussed the question of Papert's alleged advocacy of a particular pedagogic stance:

> Our experience has been that the pedagogy of discovery learning (learning without curriculum) for Logo programming mapped out in Papert, 1980 is taken quite seriously by teachers, and that little direct instruction of computational concepts, programming methods or concepts of a theory of problem solving is offered when Logo programming is undertaken in school settings. We see it at its extreme when teachers talk excitedly about how Logo can be used in 'stand alone' centres within classrooms.
>
> (PS, p. 4)

Reading this, it is hard to escape the view that the researchers have something of an educational axe to grind, a view reinforced when we subject their views on pedagogy to more critical analysis in the next section. In the meantime, we should comment on the dialectics of the situation: a single idea – that of 'learning without curriculum' comes to dominate the ideology of Logo-use, and it is this idea – rather than any of the mathematical or even psychological rationales proposed – which is subject to scrutiny by the researchers. It is refreshing to hear researchers admit that their hypotheses are grounded in what they hear and see, rather than descending from the heavens. We must bear in mind that a central rationale for the research derived from how teachers viewed the innovation (or at least, how Pea and Kurland perceived the ways teachers viewed it). As we shall see in Chapter 8, these views are shaped as much by teachers as by the innovation itself.

The third rationale was the belief that 'programming' involved 'cognitive constraints' and in particular, preplanning as a necessary component. As the authors had little data upon which to draw concerning children's programming, they started with an analysis of 'expert' programmers' performance. Pea, Kurland and Hawkins (TS) are quite explicit:

> Planning was selected as our principal reference topic because both rational analysis of programming and observations of adult programmers showed that planning is manifested in programming in important ways.

> (TS, p.181)

The rationale for this is that:

> Expert programmers spend a good deal of their time in planning programme design (F. P. Brooks, 1982) and have many planning strategies available, such as problem decomposition, sub-goal generation, retrieval of known solutions, modification of similar code for related programmes, and evaluative analysis and debugging of programme components.

> (TS, p.181)

Thus the rationale is drawn entirely from programming languages other than Logo, in cultures which include professional programmers undertaking specific tasks. Why should children with no experience of computers – let alone programming – program in the manner of experts? Surely, enough was known, as we saw in Chapter 2, to suggest that students approach tasks in rather different ways than adults. Incidentally, it is also open to question whether in fact expert programmers actually behave in the way reported *when they have no clear idea of the problem solution*. It is simple to write a plan to solve a familiar problem but not so straightforward when venturing into uncharted territories! Surely in such cases even experts spend time tinkering, reacting to computer feedback as they test out their conjectures to find a way into the problem?

Pea and Kurland recognised this to some extent, claiming that the tasks they set children should be 'sensitive to both pre-planning and planning in action'. In admitting as much, they undermined the basis of their study by recognising that preplanning may not, in fact, be a necessity; yet they adhered to preplanning as the only outcome measure of their studies.

3.1.1. *The First Experiment*
A sample of students was chosen from a private school in Manhattan. Half
were 8/9 years old, half were 11/12 years old. Some children undertook Logo
programming and an experimental and control group were given a planning
task at the beginning of the Logo 'treatment' and four months later. Some
other details were less clear. In particular:

- The experimental groups were composed of 4 boys and 4 girls (PS)
  although this is unclear: in TS, it is reported that 6 boys and 6 girls
  from the programming class were videoed as they worked through the
  planning test (we will take the former figure as more likely to represent
  the size of the sample as it is clearly stated in the *Technical Report*).
- The selection of the children was not random: it was chosen for greater
  Logo experience and higher reflectiveness.
- A control group consisted of a set of children 'matched' on the basis of
  two psychological tests to measure 'the size of basic processing capacity'
  and 'cognitive style'. The measures chosen are not justified, and given
  the small size of the sample, these are crucial as any perceived differences
  (or lack of differences) could be due to individual differences rather than
  as a result of the 'different' treatments.
- There is no information on what the control group did while the Logo
  pupils programmed.

Students were asked to devise a plan to carry out 6 chores on a transparent
plexiglas map of a fictitious classroom and to find an optimal sequence for
doing this as measured by *route efficiency* in executing the plan (how far the
distance covered deviated from the 'optimal' distance as a percentage of the
'optimal' distance).

The chores could be accomplished with a minimum of 39 distinctive acts,
such as watering the plants, emptying the trash, washing tables and black-
boards. As Pea *et al.* say, 'Finding the optimal sequencing of these chore acts
is thus a challenging task' (TS, p. 184). Moreover,

> Performing each named task such as pushing in chairs, cleaning the tables,
> in whatever order is not an effective plan. Each chore must be decom-
> posed into its component acts and the parts must then be reconstructed and
> sequenced into an effective, all encompassing plan.

(TS, p. 185)

Now as we have hinted, it is doubtful that the children's programming
involved the kinds of planning that were congruent with this task. But even
if it had, the rationale for the study would be that in the mobilisation of planning skills, the students were somehow 'abstracting' a skill of planning and
'transferring' it to a different domain. How feasible is this in two 45-minute
periods per week, during which they were not told explicitly what they were
expected to do?

The two major findings from this experiment were: i. the route efficiency score significantly increased with age from first to last plan and across age groups; and ii. the Logo programming group did not differ from the controls for any plan constructed at the beginning of the school year or at the end of a school year of Logo programming (TS, p. 185 and PS, p. 15). However, even if we accept, for the moment, the validity of the design, a careful reading of the reports reveals some grounds for questioning the findings:

- Although we are told the value of p (the probability that the results occurred by accident), the papers do not give any information on the statistical tests used. Given that $n = 8$, this is a somewhat unfortunate omission.
- The samples were split by gender, but this was not taken into account in the analysis (possibly as this would have reduced still further the sample size in each cell). If there were gender/treatment interactions, the sample can not be thought of as homogeneous, and testing of means may be of limited validity.
- The raw data in Table 1 (PS, p. 16) showed that there was considerable discrepancy in scores *before* any treatment, which casts doubt on the validity of the 'matching'.
- In Table 1 of PS, there is a score of 107.3. It is difficult to see how this occurred when the 'route efficiency' is expressed as a percentage of the 'optimal' route! With such a small sample, such errors may be extremely misleading.

Apart from statistical measures, the authors also engaged in an analysis of how the students made their decisions in planning the chores. This analysis revealed no differences between the experimental and control groups. However, it is also reported that there was no relationship between the process measures and the product measures (i.e., the 'route efficiency'). Surely one would expect the opposite: if the task involved advanced decision making, it is strange that strategy had no effect on outcome. Does this not cast doubt on the validity of the product measures?

3.1.2. *The Second Experiment*

A second experiment was undertaken as the researchers were properly concerned that the students might not have recognised the first task as an opportunity to apply insights from their Logo work as it was far removed from a programming activity. This experiment differed in that:

- A new task was devised using a computer program that 'created a graphics robot programming and testing environment, or microworld' (PS, p. 24).
- The groups (n=8) were tested twice within a six-month interval.
- The choice of the experimental group was random.
- The matching of the groups was more carefully undertaken.

In this second experiment, a clock in the computer recorded the intervals between students' moves: this was taken as a measure of thinking time. The possibility that students were doing other things than thinking in the spaces between the moves was not entertained.

The rationale for the task redesign is instructive. First, the introduction of a 'computer-based microworld environment' was designed to be 'an environment similar to the programming environments with which they [the students] were familiar' (TS, p. 188): quite what was similar is not clear, except that it was 'on a computer'. Second, the software provided feedback in a form which was supposed to allow students to see their plans enacted in a 'realistic representation' of the classroom.

There were indeed surface similarities to the kinds of direct commands in Logo. The commands in the language of the so called microworld were 'walk to', 'pick up' 'put down', 'wipe off water', 'straighten up' and the names for all the objects in the classroom. But this entirely misses the point. The aspect of the Logo language which involves direct commands of this kind (such as forward and back) is precisely the part which in general does *not* require planning of any sort. Planning *might* be useful when developing procedures, working with conditionals, the writing of functions. But this is precisely what was missing from the new task environment. More importantly, it is also exactly this element of Logo which might in certain contexts encourage children into thinking about their programs in advance, to pre-plan them. In the tasks, on the other hand, all the commands were entered one at a time, *not by the student*, but by an experimenter who typed each move onto the computer in response to the student's spoken instruction. Unsurprisingly there were no significant differences between the programmers and the non-programmers. Again, gender was a factor in the sample, and again it was not tracked in the data analysis. Thus, as in the first study, there was no evidence to support the assumption that the group could be considered as having a homogenous profile.

On page 192 of TS, there is a fascinating description of how the students behaved in the new computer test environment, which sheds indirect light on Pea and Kurland's approach to the Logo work: the children did not use the feedback aids, they did not watch the plans being enacted nor refer back to the printed copies of earlier plans.

This characterisation of students' behaviour in the tasks, calls into question what performance on the task is supposed to indicate: no transfer or – just as likely – no planning? Certainly, it is unclear what can be concluded from the researchers' observation that: 'the programming groups clearly did not use the cognitive abilities alleged to be developed in these tasks designed to tap them' (PS, p. 43). Stranger still, we read that the results were surprising in that the Logo students did not exploit the feedback available which 'because of their planning experience, programmers might have used to greater advantage' (TS, p. 192). What planning experience?

This weakness is compounded by the elaborate system for coding the response of the students, and in particular, for looking at clusters of strategies in the classroom chore task. It would appear that these analytical devices bear little if any relation to what might have been predicted, had there been any serious attempt to link the Logo activities with the problem tasks. For example, if children had been encouraged to write Logo procedures as a way of clustering particular actions on the turtle which formed a coherent sub-unit of a task, then it might have been interesting to look at the particular ways this was done, or to see whether, say, procedure names chosen in the test showed any differences between groups. This was ruled out by the design of the tasks: there was a pre-existing language for specifying the tasks.

Perhaps the most fundamental omission in both versions of the study is the lack of serious process data. This calls into question the way in which the results are taken as generalisable – conclusions of a *teaching experiment* are generalised without careful and systematic analysis of the relationship of the results to the processes of that teaching experiment. We have no process data as to *what the students actually did* with Logo. We have no evidence whether they were encouraged to plan or to reflect on plans, or whether they engaged in planning at all (we know they were told about planning in the second experiment, but that is all). It may well be that the projects in which students became engaged did not 'require' planning at all: in fact this is hinted at – though not explicitly – in the final discussion of PS, where it is suggested that students may not have used certain approaches simply because Logo may have allowed them to avoid them. It is puzzling that this is merely put forward as a suggestion after the event; surely the researchers had information which could have settled this conjecture one way or the other?

3.2. *Methodologies and the Papert-Pea debate*

At the second Logo conference at MIT in 1985, Seymour Papert launched a counter-attack on Pea and Kurland, subsequently published in Papert (1987). This remains, to our knowledge, the only serious critique of Pea and Kurland's work.

Papert's critique boils down to four points:
- Pea and Kurland were 'technocentric' – that is, they treated 'Logo' as an entity which could form a treatment while keeping everything else fixed.
- In the treatment model, negative results (no significant effects) are difficult to interpret since they only show what the experimenter failed to achieve.
- The idea that programming involved preplanning was derived from a cultural stereotype of what programming was about. On the contrary, some students find, using Logo, 'a liberation from a style that distorts their natural [unplanned - RN/CH] way of being' (*ibid.* p. 25).

- By only looking for one (an unreasonable) phenomenon, the research was fundamentally flawed since it stood no chance of discovering features of the culture 'that happen to have in common the presence of a computer and the Logo language' (*ibid.* p. 27).

We do not intend to reiterate Papert's arguments. There are, however, two points which do need elaboration. First, in his response to Papert, Pea argued that he had seriously pointed to the influence of the activities in the teaching and cultural environment, not 'programming-in-general' (TS, p. 6). Yet his evidence for so doing was based on 'appropriate provisos on cultural context for interpreting our experimental results in Logo studies' (TS, p. 6) – the provisos were mobilised at the interpretation stage and did not inform the design of the research itself. More precisely, the criticism levelled against Pea and Kurland was that they failed to link their treatment methodology with the specificities of what the children did. Although not made explicit by Papert, this does seem to us a flaw which can be legitimately deemed as fatal.

Second, Pea took Papert to task for his sweeping criticism of treatment methodologies *in general*. Indeed, Papert's critique was made rather less convincing by his strong hint that some treatment methodologies were less flawed than others (specifically, one undertaken at Kent State University which found some positive outcomes for the Logo students). Unfortunately, Pea's rebuttal was at least as unconvincing. He pointed to sophisticated instructional experiments which could be done:

> One can also do multivariate instructional experiments in which many cultural features change without 'keeping everything but one factor' constant. The logic of such multivariate methodologies, widely used in sociology and educational research, involves comparing performances of different groups to test for complex links between priori experiences, including those of social change, and changes in individuals' performances or beliefs...

> (*ibid.* p. 6)

Pea is right. Indeed, we employed similarly sophisticated analyses in Chapters 4 and 6. Where Pea comes unstuck is simply *that he did not employ any of the methods he advocated*. On the contrary, we have seen that the reports of his findings were statistically vague and methodologically unsound in a number of respects. Becker (1987) pointed out that the methodology employed suffers from familiar defects 'although no more so than most empirical research'. This may well be so, but does not absolve the researchers from seriously addressing the issues they have set themselves, particularly as Pea so convincingly documents the range of sophisticated statistical techniques he *could* have used. Instead of multivariate methodologies, we have comparisons of group means, no data on individuals' performances, and a sample size of eight children.

3.3. *Transformations of meaning: what is Logo pedagogy?*

So far, the reader might be forgiven for wondering why we are reporting a piece of research which is methodologically suspect, and whose conclusions are unsound. We reiterate that our interest centres on the studies' impact: the furore which greeted its publication; its continued citation in research findings and research proposals; its appeal to a wide range of educators outside the academic community, including school principals, education inspectors, and politicians. Pea and Kurland's work is a standard reference which is wheeled out to 'prove' that Logo does not 'deliver' by all these groups and more.[12] In order to understand why, we must look at the relationship between the research findings and the pedagogical presuppositions on which they were based.

Pea and Kurland's chosen pedagogical approach was based on their reading of Papert's position:

> Papert (1972a, 1980) has been an outspoken advocate of the Piagetian account of knowledge acquisition through self guided problem-solving experiences, and has extensively influenced conceptions of the benefits of learning programming through 'a process that takes place without deliberate or organised teaching' (Papert, 1980, p. 8).

(CE, p. 152)

Fortunately, the researchers provide the relevant page reference in *Mindstorms*. On the very same page, we find Papert clarifying what he means:

> If, as I have stressed here, the model of successful learning (i.e. Piaget etc.) is the way a child learns to talk, a process that takes place without deliberate and organised teaching, the goal set is very different. I see the classroom as an artificial and inefficient learning environment that society has been forced to invent because its informal environments fail in certain essential learning domains, such as writing or grammar or school maths. I believe that the computer presence will enable us to so modify the learning environment *outside the classroom* that much if not all the knowledge schools presently try to teach with such pain and expense and such limited success will be learnt as the child learns to talk painlessly, successfully and without organised instruction.

(Papert, 1980, p. 8; emphasis added)

Now this is a rather clear vision. Society is impoverished with respect to certain kinds of learning. The computer will enable that environment *outside the classroom* to be enriched in such a way that schools will no longer need to waste time on these kinds of learning. Moreover, like almost all learning outside the classroom, it will be achieved *without organised instruction*. The point is that Pea and Kurland have recast Papert's description of Piagetian learning outside the classroom as a characterisation of his preferred learning strategy inside school.

Perhaps this is characteristic only of the quotation on page 8? But on page 31 of *Mindstorms*, we find the following:

> ... the turtle can help in the teaching of traditional curriculum, but I have thought of it as a vehicle for Piagetian learning which to me is *learning without curriculum.*

In other words, the turtle is a medium, a 'vehicle' for learning in the Piagetian style so that, for example, the learning of other subject domains could be effected more spontaneously. Perhaps this quotation is capable of other interpretations, closer to those of Pea, Kurland and Hawkins'. Perhaps so. However, on the next page Papert states:

> But teaching without curriculum does not mean spontaneous, free-form classrooms or simply 'leaving the child alone' it means supporting children as they build their own intellectual structures with materials drawn from the surrounding culture. In this model educational intervention means changing the culture, planting new constructive elements in it and eliminating noxious ones.

> (Papert, 1980, p. 32)

It seems clear to us that Papert is quite unambiguous on this score. Yet the researchers chose to 'test' *their* reading of his position, and this led to a methodology in which some version of 'Papert's discovery method' underpinned the entire study: an essential ingredient involved 'looking more closely at the distinction between the cognitive skills that can be practised through some uses of formally elegant symbol systems such as Logo and the ways that these systems evoke particular practices in the classroom' (TS, p. 159).

This is strange indeed. Pea and Kurland appear to assume that cognitive skills 'are practised' by using Logo, and moreover, that Logo 'evokes' particular forms of classroom practice. There is little if any recognition of other forces which might shape this practice, such as those we will discuss in the next chapter.

However, it is not true that Pea and Kurland and their collaborators ignored the specificities of classroom practices altogether, although these are not available for scrutiny. In the first study, the role of the teachers was 'principally that of constructively responding to students' questions and problems as they arose' (TS, p. 180). And in the second study, the teachers decided to take a more directive role in guiding their students explorations of Logo, which in practice meant: 'the teacher of the younger class gave weekly group lessons to introduce key computational concepts and techniques, and to demonstrate how they function in computer programs. The older students were given lessons on 'Logo concepts' and programming methods 'such as pre-planning'.

So in year one, the students appear to have been told little if anything. In year two, they were told about Logo/programming concepts. Further, they were taught Logo in the style of Pascal – where pre-planning is a 'programming method'.[13] Here again, Pea and Kurland ignored the interaction between school and their version of the Logo innovation: Logo was taken as 'programming', which they attempted to move as an invariant entity into the classroom, with Papert's alleged pedagogy tightly linked to it.

3.4. The creation of conventional wisdom

Perhaps the most influential paper of Pea's group is not a research paper at all (CE: Pea & Kurland, 1984b): it is a survey of various literatures, liberally interspersed with generalisations. The key part of this paper is a section entitled 'Evidence for Cognitive Effects of Programming', and it is this – we must surmise – which became the focus of attention when the paper was published. It began as follows:

> We now return to evidence for the claims for broad cognitive impacts of programming experience, with greater awareness of the complexities of learning to program and issues of transfer. In sum, there is little evidence for these claims.

(CE, p. 171)

The section began with an important caveat: a recognition that 'studies so far, including our own, have an important limitation'. All studies cited (recall that this paper was written, we must assume, some time in 1983), looked at what Pea and Kurland designate as 'high-level or cognitive-transfer outcomes', which might only be expected to occur with experienced programmers. As the authors point out, six weeks was a typical time for 'treatments' cited, with a year the maximum duration (once a week in school lessons). In other words, say Pea and Kurland:

> ...there has been a mismatch of 'treatment' and transfer assessments because of a failure to investigate different kinds of transfer and their likely linkage to different levels of programming skill.[14]

(CE, p. 172)

Now this is a fairly damning state of affairs. The authors elaborate a careful analysis of what programming is, levels of programming 'skill development', and the 'cognitive constraints' on learning to program, but they freely admitted that little – if any – of this literature was particularly pertinent to the research question at hand. Nonetheless, they set about using what limited literature did exist for making several sweeping generalisations about programming. Here is the first:

> *Conclusion 1:* There are no substantial studies to support the claim that programming promotes mathematical rigor.

The one study cited is from Jim Howe and his colleagues who conducted a range of studies throughout the late nineteen-seventies. Pea and Kurland noted that Howe et al. (1979) reported mixed results (increase in performance on a 'Basic Math' test, and worse performance on a 'Math Attainment Test' compared to a control group), but nonetheless concluded that this did 'not support the 'rigor' claim'. Of course they are right: the claim that learning to program 'promotes mathematical rigor' was one proposed by Howe and his colleagues, and – as far as we know – only by Howe and his colleagues. Not surprisingly, they were the only ones to investigate the hypothesis in this form, and their results were mixed. In addition, they can be criticised for the selection of their sample (from the bottom two streams of ability), choice of school (a private boys' school – Pea and Kurland note this), and the choice of tests (it is far from clear that the tests measured appropriate variables).

> *Conclusion 2:* There are no reports demonstrating that programming aids children's mathematical exploration.

Now this is patently false, even at the time it was written. Pea and Kurland referred to one study using BASIC and three using Logo, and admitted that the latter 'do document children's goal-directed exploration of mathematical concepts such as 'variable' on computers'. So three out of the four studies cited reported that programming aids children's mathematical exploration. But in the next sentence, Pea and Kurland shifted the goal posts: '...studies have not shown any effects of math exploration during programming outside the programming environment'.

At the time, this was true. Papert has, as we saw earlier, always been clear that he views programming as a medium for mathematical exploration. One might surmise that the reason why extra-programming effects had not been found by the researchers cited was that they had not been looked for. More fundamentally, there continues to be little reason to expect mathematical exploration with Logo *per se* to result in transfer to performance on a pre-specified set of mathematical topics.

> *Conclusion 3:* Although Feurzeig et al. (1969) suggest that the twelve 7- to 9-year-old children to whom they taught Logo came to 'acquire a meaningful understanding of concepts like variable, function and general procedure,' they provide no evidence for the claim that programming helped the children gain insight into these mathematical concepts.

In the absence of any further comment by Pea and Kurland, it is difficult to make sense of this criticism. How would one find out if children had 'gained insight'? Is the designation of 'mathematical concepts' meant to distinguish them from variable, function and procedure in programming? Is this a methodological criticism, in that no transfer tests were undertaken? If so, this criticism is rather similar to that in Conclusion 2 above: that the 'understanding' did not 'transfer' from one context to another.

Conclusion 4: Programming has not been shown to provide a context and language that promotes problem solving beyond programming.

The evidence, according to Pea and Kurland, is based on (a) a dismissal of Papert's findings as 'anecdotes', (b) a study by Statz (1973) and (c) their own work on planning, which we reviewed above. While we do not attach particular importance to the claim itself, it is interesting to note that Statz's study (which did, in fact, claim to show evidence of 'transfer' on some but not all tasks) has itself been the subject of criticism for its methodology (Du Boulay, 1977; Weyer & Cannara, 1975), and for the dubious claim that learning Logo would somehow enhance 9–11-year olds' abilities to solve anagram/word puzzles, permutation problems, or the tower of Hanoi (see Noss, 1985).

4. THE STRUGGLE FOR MEANINGS

In this final section, we will outline a cultural framework for making sense of the ways that meanings are constructed for innovation in school mathematics, and begin to synthesise this with the largely cognitive terms in which we have framed our theory up to this point.

We begin with the response of the research community to Pea and Kurland's research. With some notable exceptions in the journals of computing education, psychology, and – to a lesser extent – mathematics education, one can still find statements that we might caricature by this (fictitious) abstract:

> This paper reports a comparative study of children using computers for the learning of arithmetic. It confirms that Logo does not produce significant enhancement of performance (Pea and Kurland, 1985), and concludes that computerised drill-and-practice exercises, within a constructivist framework, led to significant gains.[15]

It is interesting to compare this kind of invocation of Pea and Kurland's work, with their own position. Unable to locate any significant effects of the Logo 'treatment', they devote the last four pages of TS to some conjectures about what might be happening. Their conclusions are as follows:

> We can consider programming not as a *given* whose features we know by virtue of how adults do it at its best, but rather as 'a set of *practices* that emerge in a complex goal directed to a cultural framework'.
>
> (TS, p. 195: emphases in original)

> A functional approach to programming recognises that we need to create a *culture* in which students, peers and teachers talk about thinking skills and display them aloud for others to share and learn from, and that builds bridges for thinking about other domains of school and life.
>
> (TS, p. 196: emphasis in original)

...we must first recognise that we are visitors in a strange world – at the fringe of creating a culture of education that takes for granted the usefulness of the problem-solving tools provided by computers...

(TS, p. 196)

There is little to disagree with in any of these conclusions. On the contrary, Pea and Kurland have nicely formulated the need to account for the cultural setting in which activities are located, the necessity of careful pedagogical intervention, and the degree of humility which should be accorded to the investigation of new and complex educational situations. These words counsel strongly against the kind of global assertion in our fictitious abstract, and certainly do not readily lend themselves to the role which Pea and Kurland's findings have been called upon to play.

Yet in the years since they were written, we are unable to recall seeing a reference to any of them. We believe we are safe in assuming that there are hundreds of citations of the original studies, which take the findings as conclusive; yet few, if any, references to Pea and Kurland's measured and balanced conclusions of the kind we have cited above. The question is, how did one set of conclusions survive, while the other disappeared without trace?

The reasons are complex, but at root they are the outcome of a struggle for meanings. The presuppositions which shaped Pea and Kurland's work were derived from multiple cultural sources: a *technical* culture which structured their view of programming; a *psychological* culture in which the treatment and transfer methodology was still privileged; and an *educational-research* culture which was suspicious of Logo's popularity, and which delineated the notion of 'Logo pedagogy'. The truth values of these meanings are not at issue; each emerged as a factor from the web of meanings within its respective culture, and each played a part in defining the approach of the researchers.

We saw similar processes at work in our brief historical surveys of Logo in various countries. In each case, ideas shaped by diverse cultures – mathematical, educational, technical – played their part in bringing together sets of meanings for the communities involved; the enthusiasm and then rejection of Logo in U.S.A. schools, the theoretical debate in Germany, the varied and more complex state of Logo in the U.K.

In both the sketches of Logo in schools and the Pea and Kurland studies, we have seen how these cultural resonances acted back on the innovation itself, creating new meanings for the 'the innovation', not just for its 'implementation'. So, for example, just as the U.K. experience reconstructed Logo in its own image, the research community reconstructed Logo to answer *its* concerns. It turns out that Logo does *not* lead to generalised problem-solving enhancement – and ten years later, few would expect it to. But at the time, this was the question that needed answering: as we saw in Chapter 2, problem-solving was *the* current panacea to set right all that was wrong with American mathematics education.

If Logo was to be tested, it had to be against the yardstick which filtered down from the culture of broader socio-educational concerns, not the criteria of a handful of academics from MIT. In the process of testing it became changed, reconstructed in the image of these broader concerns, and ultimately found wanting. It is these meanings that linger in the minds and activities of many researchers. There is no turning back the clock: it was and is too late to restate the meanings of an innovation which had already been shown to fail in the terms laid down for it.

These meanings are not arbitrary, they are shaped by the concerns and forces in the ambient society in which they take place. Neither are they immutable; they are in a state of continual evolution. In turn, these webs of meanings evoke meanings for teachers, learners and others which are derived from, and act as, cultural symbols. Technologies – conceived in the broadest sense – develop as the function of a complex set of social, technical, economic and political factors. Each technical artefact, piece of software, or curriculum innovation contains the crystallised traces of the negotiations, compromises and conflicts which brought them into being: they are the expression of social and cultural interests in the course of struggle (see MacKay & Gillespie, 1989).

In terms of the theoretical framework of Chapter 5, we can think of this meaning-making process as an extension of the webbing idea to the cultural domain. Connections are formed between ideas from disparate cultures, and these come together to create meanings at their intersection, within the existing cultures of individuals and communities. Ideas do not, in Mao Tse Tung's celebrated phrase, fall from the sky. Ideas are held by people: individual students, their teachers, curriculum planners.

We began this chapter with two 'statements of attainment' from our U.K. National Curriculum. The first, *Explore number patterns,* and a second, designated as four levels more 'advanced', *Explore number patterns using computer facilities or otherwise.* In these two innocent-sounding formulations there are crystallised so many symbols of educational cultures: the assumption that the computer represents advancement (specifically, four levels of advancement!); the subsuming of the computer under the term 'facilities'; the belief, encapsulated in the 'or otherwise', that the tool of expression is essentially irrelevant (for what?).

Once this expression has been reified as a statement, it comes to resonate; like Pea and Kurland's research or the opinion of a respected German professor, it can be taken up as a new starting point, a new element of the educational culture, a new piece of the cultural web of ideas which informs, and is informed by, educational practice. The implication is clear. If we want to influence the direction in which children learn mathematically, we have to be sensitive to the ways teachers locate new ideas within the panoply of social and cultural relations which make up their professional practice. If we want to change the ways that mathematics is taught, we need to understand how teachers conceive of the pedagogical scenarios they are offered.

This will take us into a murky world between the cognitive and the cultural; we will, in the next chapter, devise a methodology appropriate to the task. Our central tenet is that meanings do not exist in a vacuum, they are constructed by students, academics, politicians and – our focus in the next chapter – teachers.

NOTES

1. GCSE: General Certificate of Secondary Education, a national system of examinations for the 16+ age group.
2. Some of the discussion in this section is taken from Hoyles (1995).
3. We are indebted to Reinhard Hoelzl for his comments on the discussions of Logo in Germany, and for his translation of exerpts of Bender's paper.
4. Doug Clements (personal communication) recalls noticing suddenly that Logo had been removed from the computer network at his University (in the U.S.A.). On asking the network manager for an explanation, he was told that 'DAZ-ZLEDRAW' – a 'dazzling' MacDraw-style package – had arrived, and so he had assumed that Logo was no longer needed! For Clements' network manager, Logo was a drawing tool which could be dispensed with as soon as something else became available; he had constructed Logo as a rather clumsy paint program whose educational and mathematical characteristics were irrelevant.
5. In fact a full Logo for the BBC computer was not available until the middle of the nineteen-eighties, a full four years after its introduction into schools. By this time, BASIC and – much more importantly – its culture, had taken a significant hold: an influential minority of teachers saw the computer as coterminous with programming in BBC BASIC.
6. DART is still being sold. Indeed, many are arguing – even now – that DART is preferable to Logo as it is 'easier'.
7. We have no intention of explaining the arcane details of what this, or any other National Curriculum jargon means. Interested readers will find a critical overview in Dowling and Noss (1990). We should warn that these 'statements' and 'examples' have been subject to several revisions which we have not thought it worthwhile to track. In fact, they were soon to be replaced by the 1995 version where Logo is not mentioned at all.
8. The title nicely illustrates the reduction of geometry to identifying shapes and measuring to find patterns.
9. These teachers were working with ourselves and Lulu Healy as part of a project on the use of portable computers in schools.
10. Articles about Logo in the classroom appear regularly (see for example the U.K. mathematics teachers' journal *Micromath*). This is not of course the complete picture. Some teachers still reject Logo while others still find an approach to Logo which challenges routines and resonates with their aims and objectives.
11. A historical overview of the situation up until the middle of the nineteen-eighties is given in Noss (1985).
12. Some psychologists have produced more thoughtful responses to Pea and Kurland's work. See, for example, deCorte (1986).

13. This characterisation is ours, not Pea and Kurland's. If we have misunderstood it, it is because we were unable to find published details of what actually took place.
14. This quotation contains a serious typo in the original: the quoted version here is our presumed intention of the authors.
15. We toyed with the idea of a real abstract: but we decided that this would involve giving a reference, and thus the necessity of providing a reasoned critique.

CHAPTER 8

A WINDOW ON TEACHERS

My worst time on the course so far was when you introduced dBaseIII. It was pointless. I can't run it in my school; it's much too complicated; and I can't see any use. The kids just won't be able to do it.

Bob: Mid-course Interview – 5th. month

The best thing was definitely when we did dBaseIII. I just think it's so powerful, and there is so much you can do with it.

Bob: Post-course Interview – 11th. month

Within the boundaries of mathematics education, the teacher has come in from the cold. Not long ago, the teacher was regarded simply as a 'delivery system' where, not surprisingly, the delivery did not always arrive. A litmus test of how much the focus has shifted onto teachers, can be found in the proceedings of the International Conferences for the Psychology of Mathematics Education (PME), always a reasonable reflection of the interests and priorities of at least a part of the mathematics education community. Of the 45 papers included in the published proceedings of the third conference in 1979, all but three focused on students' understanding of mathematical concepts, and, if the teacher was mentioned at all, her role was simply to dispense facts and information, identify errors or misunderstandings, provide materials or strategies to overcome misconceptions, or promote further mathematical development.

This definition of the teacher's role stemmed in large part from the prevalent paradigm in mathematics education at that time which, as we showed in Chapter 2, tended to regard problems of learning mathematics as arising from individual misconceptions. Yet even as early as 1980, some were arguing for the crucial importance of studying what the teacher did: Bishop, for example, argued that 'the teacher was the key person in mathematics education, and research which ignored this fact stood a good chance itself of being ignored' (Bishop, 1980, p. 343). As the difficulties of interpreting student responses on the basis of individual psychology became more and more apparent, 'external' factors came into focus, and these included the role played by the teacher in the learning process.

One general theme surfaced around 'teacher effectiveness' research, which was and is influential well beyond the field of mathematics education (see, for example, Brophy, 1986a; Good, Grouws, & Ebmeier, 1983). Within this school, effective teaching is summarised unproblematically as 'teaching that is successful in meeting its objectives' (Brophy, 1986b, p. 363). Teachers are seen as clinical decision-makers rather than sense-makers. This position has been subject to some criticism, both from a constructivist viewpoint within mathematics education (Confrey, 1986) and more generally (Larsson, 1986). Our analysis is similarly critical of the assumption that teachers merely transmit immutable mathematical knowledge into students' heads. Moreover, a method of investigating teachers' roles which starts by specifying objectives in terms of outcomes, treats as unproblematic the shaping of these objectives by the teachers themselves: any study which includes teaching needs some conceptualisation of the teacher's aspirations, knowledge and beliefs.

In a comprehensive review of the literature, Thompson (1992) noted a flurry of studies in mathematics education which focused on teachers' beliefs about mathematics and mathematical teaching and learning. Drawing on her own research (Thompson, 1984), she identified a consistency in views of mathematics amongst junior high school teachers, which, she argued, suggested a 'well-integrated system'. By contrast, attempts to identify coherent sets of beliefs by means of checklists have proved spectacularly unsuccessful. After analysis of the Second International Mathematics Study teacher questionnaire for the U.S.A., Sosniak et al. (1991) claimed that:

> ... eighth-grade mathematics teachers in the US apparently teach their subject matter without a theoretically coherent point of view. They hold positions about the aims of instruction in mathematics, the role of the teacher, the nature of learning, and the nature of the subject matter itself which would seem to be logically incompatible.

> (*ibid.* p. 127)

It is unclear how to interpret these disparities; indeed, it may be that incoherence is endemic to the teacher's role.[1] Consider, for example, the question of age phase. Does a teacher of mathematics have the same role, the same beliefs, across different age groups? We think it unlikely. Similarly, the evidence suggests that there are huge variations across cultures. For example, Moreira (1991), in a comparison of attitudes to mathematics and mathematics teaching held by English and Portuguese teachers, found quite different attitude profiles in the two countries, and attributed these to the differences between educational systems and the different social contexts within which schools operate. Baki (1994) has illustrated how factors such as religious beliefs and societal views of 'modernity' in developing countries, can play crucial roles in determining what mathematics teachers believe and how they act.

An important motivation for the study teachers' beliefs and understandings, has been the concern of researchers and policy makers to explain why teachers have so often 'failed' to innovate in the prescribed way – teachers seem to be obstacles standing in the way of attempts at change. A typical finding is that of Carpenter *et al.* (1986) who reported that teachers' beliefs affected how they perceived in-service training and new curricula, and therefore influenced the relation of the 'implementation' to the intentions of the original developers.

A constructive approach to the teacher's position in the trajectory of change, is suggested by MacDonald (1991):

> It was not enough, as we discovered in the 1960's, to supply teachers with better books and packaged pedagogies, although good materials and supportive advice do matter. It is the quality of the teachers themselves and the nature of their commitment to change that determines the quality of school improvement. Teachers are, on the whole, poor implementers of other people's ideas. [...] Their understanding, their sense of responsibility, their commitment to the effective delivery of educational experience for their pupils, is significantly enhanced when they own the ideas and author the means by which ideas are translated into classroom practice.
>
> (*ibid*. p. 3)

MacDonald has a point: he underscores the importance of fostering a sense of ownership of an innovation, and he points to teacher education as a means of achieving it. The question is what does teacher education involve? If it merely means a more careful attempt to teach teachers what it is thought they ought to know, we cannot believe this would be particularly effective.

On the other hand, if a serious attempt is made to design innovation *with*, rather than *for*, teachers, it might be reasonable to expect a more positive outcome. Moreover, this approach might offer a more transparent window on the evolution of teachers' thinking, and the ways in which they reconstruct new approaches as part of their personal and professional practice.

1. INVESTIGATING TEACHERS' ATTITUDES AND INTERACTIONS

In 1986, we embarked on a three-year research study – the *Microworlds Project* – to design and evaluate mathematical microworlds.[2] The setting for the study was an intensive 30-day course, based in the University spanning an entire academic year (we will continue to use the word 'course', although it may be better to think of it as a series of workshops). In fact, the course was repeated in consecutive years: the first involved thirteen participants and the second, seven. All the participants were secondary mathematics teachers and many held positions of responsibility within their schools. Broadly, the courses offered opportunities for the participants to explore and build on a range of mathematical ideas through the medium of a few powerful computational microworlds. Activities were structured around extended computer-based projects, discussion of ongoing computer work taking place in the teachers' classrooms, and the presentation and analysis of case studies of students' work with the microworlds built by the teachers. Logo programming and the construction of Logo microworlds played a central role in the activity structures, although the focus was exclusively mathematical, and other tools (spreadsheets, databases) were involved. We cannot describe here the richness of the work which the participants produced, and the mathematical discoveries which characterised many of the sessions; we hope that some flavour of the course's atmosphere will become evident as we describe the teachers' reactions to it.

Our pedagogic strategy was based on the belief that a model for learning upon which any course is based influences the model which teachers will adopt in their classrooms. As Underhill (1991) puts it: 'didactic or transmission models used to educate teachers about constructivist learning and teaching are incongruous' (*ibid*. p. 229). We began with only a loose outline of the form and content of a course within a general framework, and we were prepared to work alongside the participants in deciding the directions it should take. The crucial methodological point is that the course did not involve us in delivering the teachers any packaged 'innovation'; rather, we and the teachers co-constructed appropriate systems and pedagogies which could be used to reflect on new mathematical approaches, and on classroom strategies to accompany them.

We worked alongside teachers with a variety of objectives, one of which we described as:

... to map out some of the ways in which the teacher-participants on the microworlds courses thought and felt about employing the computer in their mathematics teaching, how their interactions with the computer influenced (and were influenced by) their pedagogical approach, and (to a lesser extent) how they integrated the computer into their classroom practice.

(Hoyles, et al., 1991b, Vol II p. 3)

Our interests lay well beyond looking for 'treatment' effects of the course: we wanted to try to pinpoint how the new ideas we introduced interacted with the participants' approaches and beliefs. Our primary focus was to understand how these beliefs changed as a result of participation in the course, and how their evolution was shaped by the aspirations and involvement of the teachers. In short, we were interested in how the teachers appropriated the course – what they did with it – as well as observing what the course did to them. In other words, we wanted to exploit the ways teachers shaped the microworlds for their own purposes as a window on their mathematical and pedagogical beliefs.

As in any study which is basically ethnographic, we collected huge amounts of data which included semi-structured interviews (before, during and at the end of the course), examination of teachers' computer-based activities and projects, analysis of the teachers' reports of the case studies of their students, observation notes of course activities, classroom observations and follow-up data obtained from post-course questionnaires. The quantity and diversity of these data provided a rich ethnography of beliefs and practices; but it posed serious problems for analysis.

Our first thought was to construct case studies of each of our teachers. These certainly captured something of the essence of each individual, but it was difficult to read more than four or five without a sense of *déjà vu*, and simultaneously hard to gain an overall impression of them as a totality. What had we done other than simply contributing to the already huge corpus of case-study data in the field? A further difficulty arose from the tension we felt between wanting to reach conclusions (rather than simply reporting data), and our wish to maintain a non-judgmental stance towards the individual course participants.

Our solution was to develop five *caricatures* of the course participants. These caricatures emerged from clustering our case studies, and choosing a representative for each cluster, which synthesised the views, attitudes and practices of its members. These were structured around a set of dimensions, spanning issues of epistemology, pedagogy and views of the microworlds project.[3] In practice, each caricature was close to a case study of an actual participant who summed up a particular profile along these dimensions, but with two differences. First, they included comments, attitudes and views which were all authentic (e.g., quotations were all verbatim) but derived from a number of participants, not just a single one. Second, our caricatures were designed to model the complexity of the data: they do what caricatures do best, focusing attention by exaggerating some points and ignoring others, sacrificing a loss of fine-grained detail in order to highlight commonalities.[4]

Thus each caricature consisted of attitudes and behaviours put together to create a recognisable individual, they were *characterisations of clusters of participants* which allowed us to 'look for' rather than 'look at' (Walker, 1981). There is, of course, a danger that the caricatures suggest an illusory homogeneity, a misleading coherence of sets of beliefs and practices. This was not our intention, but we recognise that by highlighting salient features of teachers' attitudes and behaviours in our five caricatures, we have inevitably lost some pertinent aspects of individuals and might have ignored just those aspects which made them individual personalities.

1.1. *Mary – the frustrated idealist*

Mary had been teaching for some years. She was not well-qualified in terms of initial qualifications and lacked confidence in her own mathematical knowledge beyond elementary levels. Yet she had made efforts whilst a teacher to develop her knowledge and skills through in-service activities. She was head of a mathematics department, and had also been involved in advisory work in her Local Education Authority.

Mary saw mathematics as a creative, aesthetic subject which was 'present' everywhere. People, and children in particular, could not help 'bumping into' mathematical ideas, and indeed doing mathematics without even being aware of it. Her key challenge was to help children to enjoy thinking mathematically, to feel successful and positive about the subject: this she prioritised over success in subject performance. In her classroom, Mary placed emphasis on motivating and encouraging students, and on collaborative and investigative work. She resisted ranking children by ability, arguing that abilities were subject to change. She was convinced that children found mathematics difficult primarily because of poor teaching which failed to generate confidence in their inherent mathematical expertise.

Mary's experience of computers was limited, although she felt that used appropriately they ought to be able to 'help kids discover things for themselves'. She was critical of the drill-and-practice programs she had seen and was keen to explore alternative styles of software.

On the course, Mary began with a clear professional orientation, gauging her interest and enthusiasm according to the yardstick of classroom relevance. As the course progressed, she shifted towards more personal investment, beginning to enjoy mathematical challenges and derive pleasure from mathematical projects which were not directly applicable to her classroom practice. By experimenting and exploring ways of doing mathematics with a computer, particularly learning to program in Logo, Mary began to find what she was looking for – a medium through which she could operationalise her child-centred approach to teaching and with which she could enhance her own, and her students' enjoyment and enthusiasm for mathematics.

Mary reflected deeply on how she felt about her own learning on the course. Her enthusiasm for the computationally-based activities followed a consistent pattern: first she came to see the point of a particular activity in terms of its application in the classroom, and only then found personal value in the mathematics for its own sake. Thus, her commitment to professional accountability proved to be the entry to personal involvement. Mary reflected deeply on her students' thinking and wrote sensitive descriptions of their activities to share with other course participants. She particularly noted the 'breakthroughs' made by slow learners, which she attributed to their finding a way to learn for themselves.

There were two major ways in which Mary was influenced by her experiences on the microworlds course. First, she added new kinds of pedagogies to her repertoire: this included an explicit recognition that her previous minimalist approach (don't intervene unless the child asks for help) was inadequate in a computer context. She came to see that for computer work at least, there were other, more proactive roles she might adopt which would not necessarily undermine students' autonomy and confidence. Second, and related to her first insight, Mary began to see the particularly *mathematical* contribution that the computer might offer and expressed concern when students played endlessly with Logo to create fancy colour effects (even though they were enjoying themselves!).

1.2. *Rowena – the confident investigator*

Rowena was well-qualified mathematically, had been teaching for a number of years, and held a post of responsibility within her school for mathematics. For Rowena, mathematics involved fundamental ideas and relationships with which one could make sense of the world. This clear vision of the role of mathematics translated into an explicit pedagogy which involved helping children to 'become mathematicians'. Yet this vision was in tension with Rowena's acceptance of the constraints of school teaching, constraints which she summed up as the need for qualifications and examination success. Rowena was continually trying to resolve this dilemma by adopting a highly-structured approach, while simultaneously trying to create space for students to work as autonomous learners.

Rowena regarded students' attainments as closely linked to the quality of teaching, and she resisted ranking students by ability when asked to do so. She adopted an explicitly egalitarian framework to interpret students' performance and behaviour – for Rowena, a key objective of teaching mathematics was to empower students with understandings of mathematical ideas. She had some expertise and experience with computers (she used one at home), but she was initially sceptical of their use in the mathematics classroom. Rowena felt that there ought to be some way in which the computer could play an active role in promoting the kinds of mathematical learning to which she was committed, but was unclear as to how its potential could be tapped in practice.

On the course, Rowena quickly took up the computer as a means to explore advanced topics such as calculus, and became fascinated by the elegance of programming recursively, impressed with the expressive power she acquired. Reflecting on her own learning, she expressed the view that her personal commitment and confidence in programming was a prerequisite to working out for herself how ideas might be used in the classroom.

By reflecting on her students' thinking, Rowena began to discern a specific role for the computer in her own teaching, and how it might help students to see mathematics as she saw it – an open-ended, creative subject requiring personal commitment rather than 'a great monolith of facts'. As far as her intervention strategies were concerned, Rowena recognised that the computer allowed her to 'be herself': she commented, for example, that she 'tried to keep interventions to a minimum'. Thus Rowena constructed the computer so that it offered her a means to intervene less (in line with her preferences for teaching and learning mathematics) while not compromising her commitment to 'transmitting' mathematical knowledge (in accord with her realistic assessment of school practices).

1.3. *Denis – the controlling pragmatist*

Denis was a mathematics teacher with many years' experience, and was head of his mathematics department. He did not hold high mathematical qualifications and had no experience of computers prior to the course. However, he was clearly reluctant to admit any inexperience or weakness.

Denis's views of mathematics were startlingly broad, ranging from 'something very interesting' to 'a wonderful experience', but in fact he had little to say about mathematics as a subject outside school, and regarded school mathematics as simply '..the handling of numbers'.

Mathematics teaching for Denis was defined by the school curriculum. He expressed no personal rationale for teaching mathematics, other than that it was an essential part of schooling and he had not considered explicitly why this might be. When pressed, he emphasised the practical nature of mathematics, and the necessity for students to learn the subject for their 'everyday life'. Denis wanted children to succeed in school: indeed he evaluated any curricular development in terms of its likelihood of enhancing success. As far as the computer was concerned, he saw it as any other innovation: acceptable only in so far as it fitted into the curriculum and allowed children to be successful.

Denis viewed ability as more or less fixed, and impervious to teaching. He had little trouble ranking his children by ability, and recalled cases where children had been wrongly labelled, resulting in 'banging my head against a brick wall'. Denis classified groups of children as 'intelligent' or 'thick', yet held an equally explicit view of children as 'individuals', priding himself on taking people 'at face value and seeing the way they develop'.

This tension had implications for his practice. Denis felt that most students should not be challenged unduly; given that mathematics was so difficult for so many, the solution lay in ensuring that the material presented was 'within reach', particularly for those who were unable, through lack of ability, to attain highly in the subject. Denis had a transmission model of teaching, but ascribed any failures (in terms of unsuccessful learning outcomes) to the inadequacies of his students. He wanted students to enjoy their lessons, and saw this as arising from clear guidance and secure structures, combined with the necessity of not challenging students beyond their 'ability'.

As far as computers were concerned, Denis' motivation for joining the course was that 'one was advised to move with the times ... one has to be one step ahead these days'. He wanted to 'learn the language' of computers, admitting that a key motivation for attendance was his wish to communicate on equal terms with a junior member of his department who 'knows all about them'.

On the course, Denis appeared reluctant to participate, and had to be gently but firmly persuaded to touch the computer at all. He was often reticent about joining activities especially if they required spontaneity or exploration. However, he did show some evident enjoyment of later course activities, particularly if they allowed him an opportunity to engage with mathematics which he had not previously felt able to explore. Nevertheless, Denis was overwhelmingly concerned with the professional relevance of his course activities and would not allow himself any space for investigation. He tended to reject microworlds as not particularly relevant to his classroom work, and consistently cited SMILE programs[5] as the most interesting computer applications for the classroom as they 'fitted nicely into the lesson', and moreover, required little teacher input. In contrast, he felt that Logo would need considerable teacher investment.

The combination of his prioritisation of the professional over personal, and the utilitarian over the aesthetic, combined to make Denis resistant to reflecting on himself as a learner, or even on students as learners – he often switched off when others were presenting their students' work or their classroom observations. In the classroom, Denis would not contemplate introducing any computational or mathematical idea over which he was not completely in command, but once he was – or felt he was – he wasted no time in introducing it to all his classes. Not surprisingly this did not always reap the results he hoped for, and by the end of the course, he had begun to reflect on the wisdom of pushing children round a computer curriculum without integrating the work with other activities.

Denis had a set of imperatives: that mathematics was 'needed'; that the kids were generally 'too dim'; that he must be 'one step ahead'. In reality, he found little on the course that meshed with these imperatives. On the contrary, he became rather frustrated with the lack of tight guidance, our 'failure' to prescribe classroom strategies, and the exploratory nature of the activities.

1.4. *Fiona – the anxious traditionalist*

Fiona did not have high mathematical qualifications, but she had been teaching for several years in a school where she held some responsibility for mathematics. She did not think of herself as a mathematician in any sense, and when discussing her views on the subject, often seemed to be searching for an orthodoxy which was not her own ('problem solving', 'learning by doing'). Her anxiety towards mathematics expressed itself readily – she thought it an important part of mathematical activity not to be scared – and this she combined with a view of mathematics teaching primarily defined by the priorities of the curriculum.

The pressures being applied to her in school (e.g. changes towards more investigative work) caused Fiona anxiety as she had always thought of teaching as transmitting information. She was, however, willing to 'have a go' at it – provided she could find ways of shaping it to fit her established practice and her (largely implicit) aims for mathematics teaching.

Fiona was basically pessimistic about students' competence, referring to some as 'dim' and attributing success and failure to 'intelligence'. Poorly-attaining students were 'very unintelligent' and unable to learn effectively due to their backgrounds and behaviour. Fiona saw coping in the classroom as an overwhelmingly disciplinary problem with 'keeping the lid' on the potentially explosive character of some of her students as her major priority – certainly far more important than teaching them mathematics!

As far as computers were concerned, she described herself initially as 'totally computer illiterate' and, not surprisingly, had little notion of how computers might be used inside (or outside) the classroom. On the other hand, she *was* attending the course, and she was prepared to try to come to terms with the technology, particularly for her students: 'it'll spark interest because it's something different'.

On the course, Fiona's anxiety towards mathematics manifested itself in her reluctance to become involved in any extended project and to talk about her activities. When she could see a straightforward application in the classroom, Fiona was willing to engage with the microworlds. Yet her confidence and competence grew, and she began to recognise that the computer might offer her a means to undertake and enjoy mathematical activities.

Fiona used the course as an opportunity to reflect on her students, but almost entirely from the standpoint of their emotional and affective response to the work, rather than their understandings of the mathematics. She came to see the computer as a way of building students' confidence, and this seemed to derive largely from her reflection on *her own* learning on the course ... 'it's put me right back to the position of being a student'. At the same time, she interpreted children's (and her own?) problems with computational ideas within her existing framework: 'It's made me aware of how low ability my kids are'.

Fiona adopted a traditional approach in the classroom based on considering students *en masse* as 'requiring intervention' or not. With computational activities, she decided that there was a case for less intervention on her part, and perhaps less need to 'get them on' rather than letting them 'make their own way'. By the end of the course, Fiona found herself in a dilemma. On the one hand, she felt that her 'low-ability' students needed constant intervention and highly-structured activities at all times; on the other, they needed more of a free hand, as she had seen how this student autonomy might be operationalised within a computational context.

1.5. *Bob – the curriculum deliverer*

Bob was a mathematics graduate who thought of himself as a mathematician: he had been through the highly-specialised mathematical system of a U.K. university. He had ten years' teaching experience and was deputy head of his mathematics department in a comprehensive school. He came to the course cheerfully if a little resignedly – Bob had seen it all before and did not really expect to be too much perturbed by events on the course. Nevertheless, he was interested to see 'what's new' and adopted an expectant if not over-enthusiastic stance.

Bob was not altogether articulate about his views of mathematics, despite (or because of?) his degree in the subject. He enjoyed mathematics, especially its power as a 'cool efficient means of solving problems' and was convinced it was something everyone should learn. As far as teaching was concerned, Bob was content to subordinate any personal preferences to the demands of the school curriculum. His highest priority was examination success, and his means of achieving this was to make mathematics appear 'real' to students.

For Bob, children's success in mathematics was a litmus test of 'intelligence' and determined by innate factors. Low-attaining students needed particular help, and should be protected from an over-demanding curriculum which they could not hope to master: he was concerned to find ways to help them enjoy learning. Bob saw the activity of teaching as a remedial process, helping to cope with the difficulties which would inevitably arise from 'unintelligent students' being asked to learn a complex subject.

His experience of learning BASIC during his degree had left Bob with an agnostic view of computers. In a classroom context, he saw an important role for 'courseware' which would satisfy some of the demands for new approaches being made in connection with GCSE[6] coursework. On the course, Bob took advantage of the opportunity to explore new mathematical ideas – something he had not done for many years. He was impressed with the computer as a way to solve specified problems, and this led him to a deep personal engagement with the microworlds. He came to relish exploration leading to control over mathematics and the machine; but professionally, he wanted courseware with only limited space for student control and autonomy.

Reflecting on his own learning on the course, Bob expressed the strong view that it was unreasonable that he should be expected to play with any idea before he understood it – an explicit rejection of the microworld approach. At first, he resisted engagement with activities which he could see had no classroom outlet. In fact he softened this view as the course progressed, even to the extent of citing his work with dBaseIII as his best time during the year even though he had previously rejected it in scathing terms as having no professional utility or personal interest (it is Bob's change of heart that we cite at the beginning of this chapter). As far as his students were concerned, Bob's interest was less at the level of their understandings than on what they produced; he offered plenty of examples of impressive screen displays, but little evidence that he had thought about what children might have learned. Yet his experience introducing Logo into his classroom began to perturb him as he discovered (and reported) instances of children he had thought 'less able' who were highly successful using the computer, and others who were 'mathematically good' who did not 'respond'. Bob did not reach the stage of re-evaluating his unitary concept of ability or his belief that students could not use ideas before they 'understood' them; but he found enough counter-examples in his use of the computer to begin to upset his existing (implicit) theory of teaching and learning.

His computer experience in school encouraged Bob to rethink some aspects of his teaching approach, particularly in relation to his strongly-held belief that students needed to be guided (perhaps pushed would be a better description) around a predetermined curriculum. Bob reflected on his idea of curriculum delivery, and by the end of the course had come to see computational activities as 'a legitimate part of the curriculum', in which the pressure to push students along was somewhat abated. He also saw the computer as in some way determining curriculum presentation: just as textbooks *meant* whole-class teaching, so Logo *meant* individual or pair interactions. It was sufficient to present the right tasks in the right way: what students learn was, in any case, governed by their innate ability.

2. TEACHERS MAKING MEANINGS

We will take the liberty of thinking of Mary, Rowena and the rest as real people, even though they are essentially collages of individuals. Our justification is twofold. First, when we showed the caricatures to some of the teachers involved, they recognised each other, and – with surprising good humour – themselves! Second, we have consistently found that the caricatures resonate with teachers who have perused them, in settings such as in-service sessions.

If little else, the caricatures illustrate the complexity of interaction between epistemological, pedagogical and professional considerations as the teachers made sense of new ideas. As they struggled to make these ideas their own, there is no simple sense in which we can describe their straightforward adoption or rejection of them. The ideas were shaped in interactions with computers, colleagues, and back in the classroom: it is clearly fruitless to seek straightforward determinants of teachers' beliefs.

We will separate our discussion into three related parts, focusing respectively on school mathematics, the microwolds innovation, and pedagogy. Each will centre on the reshaping of different aspects of the innovation, although they do, of course, overlap to a considerable extent.

2.1. *Reshaping school mathematics*

Rowena and Mary provide a helpful starting point. Rowena's explorations with the computer connected with her mathematical focus, providing a setting which linked to her primary interests as a mathematician. Reflecting on this experience gave Rowena an opportunity to rethink her pedagogical ideas; from a secure, familiar and newly-exciting, mathematical base, Rowena could begin to resolve a pedagogical dilemma with which she had been tacitly struggling for some time.

Mary's trajectory was almost the reverse. Her starting point was pedagogical, stemming from her deeply held convictions concerning child-centred pedagogy, and discovery learning. Mary's implicit belief was that these approaches were incompatible with a mathematical orientation; to be true to her pedagogy meant sacrificing mathematics, an endeavour in which she herself was not confident. But Mary discovered that she could do mathematics, and moreover, that she could do it in just the way she hoped for her children. She began to see that discovery and enjoyment were not antithetical to mathematical activity, for the first time connecting mathematics with 'feeling OK'.

Rowena and Mary made meanings of their new experiences in different ways, connecting the mathematical and the pedagogical, webbing their personal and professional knowledge in opposite directions. Where such connections were not made, as we saw in Denis's case, it became very difficult to review either the personal or professional relationship with mathematical activity. We (ourselves and the other teachers on the course) simply did not provide Denis with a familiar foothold he could use to connect with the new ideas he encountered.

Bob's mission was delivery rather than innovation. To what extent was this role imposed on him? To some extent certainly, given the practical day-to-day demands of teaching and the fact that the culture of Bob's school encouraged rather rigid ways of thinking about curricula, students and criteria for success. On the other hand, Bob had a tacit theory of school mathematics and it is this theory which he operationalised in his classroom. For example, Bob initially believed that one way to overcome the 'natural' aversion of children to mathematics was to introduce 'reality' into the classroom. It seemed that the combination of his own expertise in the subject, the perceived difficulties his students experienced and the suppression of his personal approach in favour of 'the maths scheme', conspired to produce a view that 'reality' was the missing link – a solution to the problem that 'children don't see the point of mathematics'. It is worth noticing that in his own mathematical background, this is precisely *not* what attracted him to mathematics: what attracted Bob was its power and abstractness. So Bob constructed his own meanings for school mathematics: he saw himself adopting strategies to overcome his students' difficulties, and he added to his repertoire some of the ideas he encountered on the course.

We cannot predict with certainty how meanings shift. Why, for example, did Mary develop new mathematical insights which went well beyond the classroom while Fiona did not? The answer seems to lie in their respective feeling for or sensitivity to mathematics which, in Mary's case, was certainly present before the course, and which the course developed. As we have seen, Mary, unlike Fiona, developed a line of activity which began with the personal. She allowed herself to forget her students, and in so doing, discovered mathematical territory she would never have visited if she had focused exclusively on their needs. Fiona's priorities centred firmly on the school: she wanted to find ways of teaching more effectively, and she never relaxed this agenda.

The course provided Mary with the realisation of her existing ideas about mathematics – it *was* enjoyable, and above all, she *could* do it successfully. It was this personal expression which, we surmise, brought about the shift in her pedagogical views: the computer in a sense liberated her to see that enjoyment in the process of doing mathematics could be realised, not only for her, but for her students: that there were new kinds of mathematics, new domains of abstraction which were learnable, enjoyable *and* legitimate. Fiona, on the other hand, shaped all her microworld activity to the demands of the curriculum, and in so doing did not allow herself to activate any connections to new kinds of mathematical knowledge.

2.2. *Reshaping the microworlds innovation*

The computer seems to have played an important role in shaping the meanings the teachers constructed. Most obviously, everyone but Denis came to re-evaluate the computer's role in their professional or personal practice. This is hardly surprising, given that the computer played such a central role in all the workshop sessions; but we should not ignore the anecdotal evidence of many shorter, more technically-oriented courses, where such shifts in attitude do not occur and which sometimes engender rejection of the computer.

But this is where the homogeneity ends: how the computer was constructed varied across the caricatures. Let us start with Fiona, the anxious traditionalist. She was firmly committed to a view of how mathematics must be learned: it must be unthreatening, it must be achievable, and it must be well within children's inherent capabilities. She was initially anxious that the computer would be unable to provide this kind of support, and saw no way in which it might form a legitimate part of her practice. As the course unfolded, Fiona began to exploit the computer activities as a way to build her own mathematical confidence, concentrating exclusively on mathematics within the given curriculum, yet finding in her explorations some liberation in being able to ignore the difficulties of her students (difficulties which were a projection of her own). In building her own confidence, she began to see ways she might project this back onto her students.

In the case of Bob, we see an interesting dialectic at work. By the end of his activities on the course, Bob saw, for the first time, that the computer might play a legitimate role in his mathematics teaching. But this legitimation was double-sided. On the one hand, Bob seemed content to browse the course searching for aspects which would allow him to cope with new professional pressures. On the other, he had to make these approaches his own and find ways of working with the computer which he could incorporate into his existing practices. This legitimation was not a passive process: it was not as if Bob simply picked up ideas which he could use while rejecting those he could not. In the process of working with the ideas of the course, Bob reconstructed the ideas in ways which enabled him to feel comfortable. It was in the process of reconstruction that the match between his imperatives and those on offer was made or rejected, simultaneously shaping and connecting the new mathematical and pedagogical ideas he encountered.

Denis, like all counterexamples, provides food for thought. Denis had a series of imperatives for his practice: children are mostly not capable of doing mathematics, they are individuals who need to be nurtured with 'success', the curriculum is a given. On a personal level, there were imperatives too: he could not conceive of using an idea before he thoroughly understood it, and he was impatient with exploring for himself when he knew full well that the answer was available. Denis's experiences with the computer changed few, if any, of these beliefs; his level of engagement with the computer never became deep enough to catalyse any reassessement.

So far, we have seen how the computer provided a resource which could be used to web meanings for the innovatory practices and pedagogies we introduced. But as we stated at the outset, we saw the locus of change as located primarily in the redefinition of what counts as mathematical.

In the cases of Rowena and Bob, both felt a confidence with mathematical ideas which allowed them to shape the computer work mathematically and develop a critical and reflective stance towards the microworlds from a mathematical perspective. But if we look more closely at Rowena and Bob, we find two quite different mechanisms at work. Rowena had a flexible view of both students and the curriculum, and found in the computer a means to resolve the tension between her views of what teaching should be like and how it actually was. Bob, on the other hand, was more curriculum-centred, more reactive, but became more flexible in his views of teaching and learning; paradoxically, this flexibility emerged from his letting go of his pedagogical concerns, and focusing on personal, mathematical exploration. For both, a key element was the enlargement of the mathematical domain provided by their computational activities.

2.3. *Reshaping pedagogy*

Before we leave our teachers, we should consider the question of pedagogy a little more closely, particularly as it will help us make sense of the situation we encounter in school in the following chapter. Again, we find very different trajectories of change.

Mary provides an example of someone who initially interpreted the Logo philosophy as 'don't intervene' – a common enough reading among teachers (at least in the U.K.) and, as we saw in the previous chapter, among researchers as well. In the course of her activities, Mary came to replace her position of 'minimalist intervention', with one which was considerably more sophisticated: she began to reflect on how her role as a teacher might change as a result of the computer's introduction. Mary's motivation was precisely that she wanted to focus the children's attention on *mathematical* ideas. Recall that Mary started off by arguing that 'maths was everywhere' and that 'children should discover for themselves'. Yet here is Mary aware that mathematics is more than an accidental process of discovery, and intervening consciously to maximise the chances of her children connecting to mathematics through their computer-based work. Once more, there is an epistemological as well as pedagogical connection.

For Rowena, a different mechanism was at work. She began the course with a strongly held set of beliefs about mathematical investigations, centred around the primacy of *doing* above telling: yet she was aware that she spent 'too much time' directing students' work. Rowena found a means by which she could relieve the tension between her theory and her practice. She found a mode of intervention (less 'telling' and more contingent guidance) through which she could prioritise *both* her preference for an investigative approach *and* her focus on helping students to master mathematical content. Unlike Mary, Rowena's trajectory of change lay in the direction personal → professional, whereas Mary's was professional → personal.

3. CONNECTIONS AND CULTURES

Our conclusions are threefold. First, epistemology matters. The teachers who found new mathematics, used this as a basis for reflecting on their pedagogic concerns. In many cases, this involved a broadening of the range of activities which counted as mathematical, the inclusions of domains of abstraction which had not previously been considered.

Second, the computer was a resource for webbing teachers' ideas and beliefs. It catalysed connections between personal, pedagogical and professional beliefs, allowing participants diverse ways to shape innovatory ideas to their own dispositions and predilections.

Third, the computer offered teachers a window onto their own roles. By affording them an opportunity to step out of purely professional roles, and offering activities which formed the basis for reflection, some teachers came to adapt their view of their roles, and in some cases, construct new ones.

More broadly, we see that when we try to access the complexities of teachers' attitudes and practices, we find they are simply not susceptible to the location and importation of 'effective' ways of using computers in the classroom. Change was contingent on a shifting pattern of beliefs and ideas; what mattered was the connection between activity and personal or professional interests.

More generally, beliefs and beliefs-about-practice were dialectically related, with the direction of the trajectory of change depending on a multitude of factors: the beliefs themselves, the relationship of the innovation with what is deemed as mathematics or school mathematics. How the computer was reconstructed in each individual case highlights what we mean: Rowena came to balance her vision of mathematics teaching, a vision which had been in tension because of its apparent impracticability; Mary found a medium with which to operationalise her affective orientation; and even Fiona, who in other respects we might find conservative and unchanging, found a way to legitimate her existing practices and priorities as mathematical, by coming to view computational activity as a situational domain of abstraction, rather than merely a diverting, but non-mathematical activity.

3.1. *The social construction of change*

This picture provides some insight into the meanings of change construed by individual teachers. But it tells us little about the process of change itself, or about the relationship between personal construction and pedagogic action. Yet if we are interested in revisioning school mathematics, we will have to take account of broader, cultural concerns, preferably in actual schools and classrooms.

One approach has involved attempting to be more explicit in understanding the complex interplay between what teachers say outside and inside the classroom. If we return to the literature, we see a range of findings. For example, Thompson (1984) noted a strong, though subtle, relationship between teachers' views and classroom practice. In contrast, other work has identified mismatches between beliefs and practice referred to by Ernest (1989) as 'the espoused-enacted distinction'. In sum, the evidence suggests that the relationship between beliefs and 'beliefs-in-action' is not straightforward (Brown, Brown, Cooney, & Smith, 1982) and we are left with a research problem as to how to tease out any mutual interaction.

Another approach has involved directing more attention to the constraints of the teaching situation. For example, Desforges and Cockburn suggest that:

> We have worked with many teachers who were well informed on all these matters [children's mathematical thinking, clear objectives, attractive teaching materials – CH/RN] and yet who routinely failed to meet their own aspirations which they shared with mathematical educators. We do not believe that this failure has been due to idle or uninformed teaching. On the contrary, we suspect that the failure lies in the unwillingness of the mathematics education establishment to take seriously the complexities of the teacher's job.

> (Desforges & Cockburn, 1991, pp. 1–2)

This complexity has similarly been recognised by Arsac, Balacheff and Mante (1992) who point to the constraints of classroom culture on mathematical work, such as lack of time and teachers' feelings of responsibility for acceptable, closed kinds of student outcomes. Significantly, they also point to the influence on teachers' classroom behaviour of personal ideas about learning and teaching mathematics, although they contrast the teacher's own problem-solving ingenuity and her classroom practice. The situation is made still more complex if we take account of other teacher anxieties, deriving, for example, from anticipation of parental opposition to any new method, and the possibility that experimentation might lead to deterioration in examination results (Nolder, 1990).

Yet a number of researchers have argued that the relationship between teachers' beliefs and their operationalisation in the classroom, cannot be completely accounted for by the notion of constraint (see, for example, Goffree, 1985; Olson, 1985; Romberg, 1988), and many invaluable and detailed studies have shown how meaning is constructed reciprocally in classroom interaction.[7] This observation is key. Teachers are not pushed arbitrarily by 'constraints'. Neither are they free agents. There is a mutually constructive relationship between what teachers believe and what they do. What our caricatures indicate is that a teacher's ideas about mathematics and mathematical microworlds shapes the ways he or she operationalises these ideas in the classroom: at the same time, the cultures and practices of the classroom cannot be left at the University gate, they are constitutive of what is believed. This is only a special case of the dialectical relation between cultures and mathematical meaning-making that we have sought to understand in this and the three previous chapters.

Our explorations in this chapter illustrate that a cultural perspective is crucial. It indicates that, for teachers and children alike, the construction of meaning involves constructing webs of connections within and between cultures. Maintaining and strengthening this cultural web is a crucial element of change.

In the following chapter, we will therefore move into a more naturalistic setting, and explore the social construction of change in a school. This will allow us to recognise the diversity of practices in which teachers participate, and the heterogeneity of the discourses which govern them. We have come to think of our caricatures not as stories about people (or classes of people), but about cultures. Anxious traditionalists, curriculum deliverers and frustrated idealists are made, not born. They develop meanings from their professional practices and from cultures which lie within school and beyond. What are the factors which influence how teachers make sense of change? And how do teachers and students together construct meanings in schools for the knowledge domain known as 'school mathematics', or for 'the computer'? What can we learn from studying the complex process of change within school cultures and practices?

NOTES

1. There is a complicating methodological difficulty. Studies of teacher attitudes and beliefs have adopted widely disparate methodologies. Thus, the increasing focus on the teacher's perspective has led to a range of novel research methods which have made comparison and generalisation difficult. The substance of these arguments have appeared in Hoyles (1992).
2. The project workers consisted of ourselves, and Rosamund Sutherland. The project was funded by the Economic and Social Research Council, grant number C00232364. Part of what follows is taken from final report of the project: Noss, Sutherland & Hoyles (1991).
3. The dimensions were: views of mathematics; feelings of mathematical competence; articulated aims of teaching mathematics; views of school mathematics and its relation to mathematics 'outside school'; expectations of students' mathematical attainment and attributions of success/failure; pedagogical orientation; attitude to computer use in mathematics; focus of course activities; classroom use of computers; and views of participants' own intervention strategies.
4. The caricatures were not typifications; they did not represent 'ideal types' in Weber's sense.
5. School Mathematics Investigative Learning Experience. These mathematical programs have acquired something of a reputation for being useful, limited and directed at clearly delimited topics of the curriculum.
6. General Certificate of Secondary Education, a national examination taken at age 16 years.
7. We have only just obtained a copy of Cobb and Bauersfeld's (Cobb & Bauersfeld, 1995) collection of studies, which focuses on the issue of mathematical meaning from a perspective which is different from, but complementary to, our own.

CHAPTER 9

A WINDOW ON SCHOOLS

In the previous chapter, we tried to develop a picture of the teachers' view of change, by synthesising cognitive and cultural perspectives through the construction of our caricatures. We are aware that our analysis is still incomplete. Recent research has properly begun to respect teachers as part of a community of practitioners with common cultures developed alongside other players in the educational field. There has been a growing appreciation amongst innovators of all complexions that the unit of teacher development should be the school rather than the individual teacher. For example, MacDonald (1991) distinguished three main changes in thinking about innovation over time: from package development, through teacher development and finally school development. The idea is that teachers can do rather little on their own in the face of the weight of the collective practice of a school, its institutional habits and routines.

This recognition goes some way to recognising the complexity of educational innovation. Yet it still adopts a viewpoint which characterises change agents (in this case, the teachers) as thwarted, stymied by habits and routines. By contrast, we have attempted to see beyond this, to try to formulate a theoretical perspective with greater explanatory power, based on making sense of the interplay of meanings imported from diverse cultures.

In this chapter, we offer a case study of the introduction of Logo into a single school. Our reasons for doing so are twofold. First, we take a step nearer to the educational reality experienced by most children and teachers; those involved in schooling will, perhaps, recognise some of the struggles, tensions, and complexities of individual and cultural meanings which characterise our story. More fundamentally, the school provides a setting in which we can synthesise the perspectives of earlier chapters; the individual views of the students, the culture of innovation and innovators, and the perspectives of the teachers.

Our case study will trace the transformation of the school's original goals for the Logo work as it became appropriated into the school's practices. It will turn out that there is a strong epistemological strand to the process, and that a crucial chapter in the story concerns the different ways that the knowledge domains of Logo and mathematics were conceptualised by the participants.

1. THE BACKGROUND TO THE CASE STUDY

Valley primary school had been part of a major Logo initiative[1] in England in the early nineteen-eighties, a period in which there was still considerable freedom from prescriptive curriculum constraints in U.K. primary schools. Many of the children had made enormous strides in their appropriation of programming and of mathematics. As these students progressed into their final year in junior school, their parents began to ask questions about the computer provision which would be available upon transfer to secondary school, in particular one – Harston school – to which a cohort of students was bound the following September. It is Harston school which is the setting for the case study.[2]

Prior to 1985 there had been no institutionalised computer work in Harston before the age of 15 years, when computers were introduced as part of an optional 'computer studies' course. But as a result of pressure from several powerful parents of children at Valley, the head of Harston set wheels in motion to introduce computers lower down the school. We trace what happened over the next few years from the points of view of the head teacher, the teachers primarily concerned with the innovation (that is, the heads of mathematics and of computing), the Local Education Authority (LEA) through the voice of the advisory teacher for mathematics, and some of the students.

After much discussion, the school came up with a plan to introduce computers into mathematics classrooms in a gradual way, into two or three 'first year' mathematics classrooms in 1985/6 expanding to a further two or three second-year mathematics classrooms the following year.[3] The school chose mathematics for a variety of reasons, notably because of the support available to do this from ourselves and the LEA advisory teacher. In addition, John, the head of mathematics, although not overtly dissatisfied with the school's mathematics curriculum was predisposed to explore new ideas in teaching and learning mathematics – he was a part-time Open University tutor for a course in mathematics education. John was willing to take on the innovation particularly as Colin, the head of computing, was keen to join in. Colin, though a BASIC devotee inexperienced in Logo, was willing to learn the new language. In addition, he already taught some mathematics lower down the school (although he was, by his own admission, under-qualified to do so) and was therefore in a position to provide technical support to the mathematics team.

There were financial concerns too. There was already a prevailing shortage of machines and funds at that time, and the promise of free hardware and software to be provided by the LEA was a powerful incentive for the school to participate.

In summary, the catalysts for introducing Logo into Harston were parental pressure and the necessity to maintain or enhance local prestige; the innovation was made possible because of the provision of hardware and support, and the agreement to participate of key personnel in the school; but the mathematical focus envisaged for Logo was determined by the rather fortunate synergy of local and school factors – University and LEA support coupled with teacher willingness and administrative convenience.

The term before the entry of the 'Logo students' from Valley, John and Colin volunteered to join an evening LogoMathematics course held at the University over a period of one term (rather similar to, though much shorter than, the *Microworlds* course described in the previous chapter). They were introduced to the main ideas of Logo which they explored, alongside other mathematics teachers, within a variety of projects. During the summer, plans were made amongst the 'innovation team', which comprised ourselves, Claire (the advisory teacher from the local authority), Colin, John and one other mathematics teacher from Harston. Concerns centred on how to begin with Logo in mathematics, how to organise the Valley children together with those who arrived from other primary schools (who had no Logo background), and how to support other teachers in the mathematics department who wanted to join in. Colin decided to write a series of Logo worksheets to be used by all the staff. These focused on learning the language since he anticipated that ideas for projects to explore mathematics with Logo would best be developed as they emerged in the course of the year. As it turned out, this rarely occurred.

As far as class organisation was concerned, the team decided that it would be easier to cope with the diversity of Logo experience if the Valley children were kept together in Colin's class. This would enable other children to catch up with elementary Logo while Valley children pressed ahead. The plan was that there would be intensive 'outside' support for one year and then a gradual reduction as the teachers became more familiar with Logo in the classroom and its potential for learning mathematics. Throughout the first year, Claire and ourselves made regular visits to the school working in the mathematics classrooms, helping the children, discussing the curriculum, devising activities and advising the teachers. The following year, Claire took over responsibility for regular school support visiting on a weekly basis, while we took on a rather different role, occasionally visiting to monitor progress, act as trouble-shooters, or try out new ideas. Finally, during the year 1987/88, the school took over the innovation completely with no external support, since by that time a curriculum for the LogoMathematics work had been devised, at least for the first few terms. John and Colin continued to be the major agencies of the innovation although one other teacher was beginning to participate and was enjoying the work. Yet, by the end of the school year in 1988, the school had decided to remove Logo from mathematics and instead make it one plank (to be shared with word processing) in the school's general provision for 'IT awareness'.[4]

The public reasons for the school's change in policy were local and pragmatic: there was not adequate access to machines, Logo took away too much time from mathematics, and many teachers – in fact all of them except John and Colin – were still uncomfortable with the software. Neither can it be a coincidence that the planned change was in line with new policy directives from both national and local bodies which set out an 'entitlement of IT' for everybody; in the late nineteen-eighties, this was envisaged as taking place through IT courses, rather than through the curriculum. The content of these new courses was a major issue for schools at that time.

Clearly the goals of the Logo initiative had been radically transformed: yet the reasons offered appeared merely technical, local or policy-driven rather than educational. Was the failure of the LogoMathematics initiative purely a matter of resources, teacher inexperience and new curricular policies? Or were there other, more subtle, factors at work? In order to unravel this puzzle, we conducted a series of interviews with the major actors in the innovation; the head teacher, John, Colin, Claire and a selection of students (now 16 years old) – some of whom had been the stars of our earlier case-study work. Our aim was to analyse the introduction of Logo into the school from these different vantage points in order to tease out the main influences on the trajectory of change.

2. THE TEACHERS' VOICES

We began with the teachers. In all the interviews we adopted an unstructured interview style based around a few general questions. We wanted to leave space for the teachers to express their own opinions, to elaborate the factors they saw as leading to the change in policy direction, to provide their evaluation of the experience as a whole and their feelings about the old and the new organisation of the Logo work.

2.1. *The head teacher*

The head teacher was faced with a set of problems in 1985 which had, to a certain extent, been foisted on him by the parents from Valley. How was the school to cope with Valley students whom he wanted to attract, but who were a problem, as they were already rather sophisticated Logo users when Harston had little or no Logo experience? How could the work of the Valley students be integrated with that of other less-experienced children? How could the head coax his staff to join in the work? And how could he support the teachers and the students?

We conducted a semi-structured interview with the head with the following questions in mind:
- What was his view of the initial computer experience in September 1985?
- What were his policies in the school in relation to educational computing at that time? How and why had they now changed?

By introducing computers into mathematics in 1985, the head felt that he had solved an immediate and pressing problem. He had found a way which would allow Logo to be introduced into his school with the agreement of the heads of mathematics and computing, both of whom were crucial for the innovation to work. There would also be support, machines and training because of the commitment of the LEA and the University. But looking back on the Logo experience, the head now regarded this original policy as wrong – perhaps it was right at the time, but now he interpreted it only as an interim measure:

> ...it enabled us to have the debate (about the use of computers) at a higher level; but Logo as an 'experience' wasn't something we should have done. It enabled us to get through a growing period – we experimented but inevitably were on the wrong track at the start.

It was clear that the head had never conceived of Logo as mathematical, and he had placed Logo in mathematics more as a matter of convenience than as educational policy. The head had been sanguine about an innovation which would 'enhance general problem-solving skills' by offering a 'Logo experience', and mathematics lessons would provide as good a setting as any other. Now he was happy to shift the organisation of Logo work out of mathematics altogether, as he saw limitations in his earlier idea of Logo as an object of study, and now preferred to see it as a tool: still to assist in 'problem solving', but also as a route to 'IT competence'. Ironically enough, we interpreted the change in exactly the opposite way – we thought that Logo had initially been introduced as a tool for mathematics, and now it was viewed as an object of study in its own right.

The change in the head's view had the added advantage of ending what he had seen as a problem with the previous approach, namely the monopolisation of the school's computer resources by mathematics and moreover, by only a small proportion of the pupils:

> We gave too much priority to the Valley kids so they were overfavoured.

Clearly there was a need for the school management to balance out available resources, but underlying the new position emerged a strongly-held view which depicted computer use in schools as the delivery of general IT skills with software including Logo, augmented by software in knowledge domains which 'fitted' the curriculum – the head argued that the 'release' of the BBC computers from Logo would allow them to be used in mathematics 'functionally'. We asked him what he meant by this, and he explained that he meant 'only when needed'; he gave an example, the program VECTORS, a self-contained curriculum package which needed no teacher expertise to organise. This brought to light another, less explicitly articulated, factor underpinning the change in policy. It was practical. It circumvented the considerable teacher reluctance to use computers in their classes – the innovation had hardly spread beyond the classrooms of John and Colin – and it enabled optimal hardware allocation by having specialised computer classes for whole groups. It also had the added bonus of being defensible: content-specific software was directly related to the curriculum and easy to use, software for problem solving in IT brought the promise of flexibility linked to utility. In the eyes of the head, the previous attempt to link content- and process-related aims had rendered both problematic.

Two further issues emerged. The head had expected that the Logo experience would generate enthusiasm for computer work amongst *all* the students. He had also hoped that parents would continue to make demands on the school for more computer use in general and more Logo in particular. He was surprised and saddened that this had not happened, and he blamed weakness in the organisation of the Logo work for the apparent absence of local enthusiasm and the lack of a community of computer users. The head had hoped for a local 'presence' in the parental community, and this had not emerged; and in the light of the emerging competition between schools that characterised the early nineteen-nineties, he was deeply exercised by the threat of losing his students to the private sector. Thus the selfsame concern for image and 'keeping up his intake' which had galvanised him to start a prestigious and high-profile innovation now demanded that it be given up. In this respect, echoes from broader educational cultures clearly weighed heavily; as the head commented: 'Logo is a spent force'.

2.2. *Two teachers: Colin and John*

When we talked to Colin, the head of computing, we learned that he had spearheaded the new direction for computer use in 1987/8; that is the removal of the Logo work from the auspices of the mathematics department and its introduction into new courses of IT awareness. In fact, it transpired that he had given up using Logo (and any other computer use) in mathematics half-way through the previous year.

The main reason Colin gave for stopping Logo in his mathematics classes was practical – the students working with Logo demanded his constant attention, dominated his energy and took away from his time for mathematics:

> It's all down to class size. As soon as your back's turned the kids have problems. It is simply impossible to deal with a whole class and with six children working on three computers.

But there was more to it than this: it appears that Colin had never accepted that Logo programming was mathematical, and he had come to view it as a diversion from a mathematical agenda, defined and delivered by the topics in the curriculum. Logo had been added on to mathematics through his worksheets, but he could not see how to integrate Logo and mathematics in any natural way. If Logo was simply seen as a programming language or as 'an experience' then there was little or no justification for including it in mathematics lessons. For Colin, 'it's just another distraction...' and one that took up considerable time and energy.

But Colin's position was not altogether straightforward. He became deeply committed to Logo. After the initial scepticism frequently encountered amongst computer studies teachers weaned on BASIC, Colin – who was open-minded and keen to take on new ideas – had become a Logo convert![5] The perceived dominance of the Logo work in his mathematics classrooms was perhaps fostered by Colin's own behaviour – he may in fact have enjoyed being 'distracted' by Logo, especially as it transpired he was more secure in programming than in mathematics: recall that Colin's mathematics teaching was, after all, simply a fill-in. Colin was a 'confident investigor' in computing; but as far as mathematics was concerned, he is better described as an 'anxious traditionalist'.

The second reason Colin put forward to explain the turn of events was the lack of confidence and competence he observed when working with other teachers: this had spin-offs in terms of students' competencies;

> Only some teachers could do the computer work properly while the others did not like it and could not do it – some children at the end-of-the-year Logo test could not even put a disk in the drive!

Note that Colin assesses the students on computer skills, not on any mathematics they might have learned. Interestingly enough, he did not interpret this as his problem – even though it had originally been assumed that he would take on in-house training for the other mathematics teachers via a cascade model. In fact it transpired that this had never taken place – John had not pushed it and Colin had not volunteered. This inaction could reasonably be put down to lack of time and energy but there were clues that shortage of hardware was another factor. If more teachers joined in the innovation, the computers would no longer remain exclusively available to John and Colin. Not only would they lose access, they would be subjected to all the inconveniences of moving equipment between classrooms and monitoring its whereabouts. In this respect, it was, perhaps, in everybody's interests to circumscribe the impact of the innovation; John and Colin's as much as the other (mainly reluctant) teachers.

But Colin, as we said, had become a Logo convert; he wanted to teach Logo *per se* outside the shackles of mathematics. Thus at Colin's instigation, the innovation was not extinguished: it was hijacked from mathematics to IT. In fact Colin became the major proponent for Logo for younger children in the school, so 'pupils could see that a computer can be programmed'. Programming for him was an important way to learn how to control a computer: and to Colin's way of thinking, this was justification enough – mathematics simply did not come into the picture.

Other facets of Colin's practice changed as his enthusiasm for Logo grew: he insisted on better physical conditions for his work, the lab was expanded and made more pleasant and airy with posters on the walls, plants at the windows and the computers in groups rather than in lines to encourage social interactions. As far as the Logo curriculum was concerned, Colin no longer relied on his sequence of worksheets but allowed students to undertake projects, with the worksheets taking on the role of a glossary of commands for support. He became deeply committed to Logo, but was adamant that it's exclusion from mathematics was good for mathematics! It's 'much better for maths without the constant interference of the computers – also everybody has a chance and not just a few – and they are not dependent on inexpert staff' – note that 'expert' meant expert in computers. We have more than a suspicion that Colin also believed that mathematics was a distraction from teaching Logo rather than vice versa!

John, the head of the mathematics department, was clearly quite relieved that the responsibility for maintaining the momentum of the introduction of computers was no longer his. John too, was in the process of reconstructing for himself the meanings he brought to the innovation. Echoing the thoughts of the head, he suggested that the change to using computers in IT was an 'obvious extension' of the previous work. He realised that he had now '...lost a maths period' (they had been granted an extra mathematics period when the computer work had been introduced in 1985) but he felt that in fact some children were doing *more* maths now as they were not spending so much time with Logo.

He suggested many practical reasons why things had changed: resources could now be more equally shared; 'The whole year group now gets covered' and the Logo work would be more central in IT rather than a diversion in mathematics: 'Kids need better back-up for it to work, with some competent teachers [and] printed material'. John admitted that the Valley students had made little progress in their Logo work, putting this down to poor hardware provision. Generally, John's opinion was that using Logo required too much 'overhead' for most kids – and maybe most teachers too. Teachers in his department, he argued, had found Logo hard, and this was the reason why nothing really had been done to spread the innovation. Was that the only reason? As we probed deeper, another factor emerged: perhaps John too had found it hard to cope with the new Logo work and to develop ideas to exploit it once the first elementary stages had been completed. Inevitably therefore, he, together with some of the Valley children, had become bored:

> We got fed up in the last year as the computers were not powerful enough to do the quiz-type programs.

We wonder whether John's laying blame on the hardware might not mask a more fundamental reason. After all, quizzes clearly *could* have been programmed in Logo with little problem, but with what purpose in mind? It became apparent that John wanted the students to write programs to ask mathematical questions such as 'What is the sum of the angles in a triangle?' and to give the correct answers after several attempts. The mathematics was in the questions and answers, as defined by the topics in the syllabus, not in the program constructions. So it emerged that John, like Colin, could only conceive a mathematical dimension to programming in so far as it dealt with 'mathematical' questions, i.e., clearly specified topics of the curriculum. His explanation of why the Logo work had not been integrated into mathematics lessons was Logo's fault – it did not 'fit'. John, like Bob in Chapter 7, was a 'curriculum deliverer'. As such, his choice for computer work was inevitable and logical: to dispense with Logo altogether and to use CAL software which closely fitted the curriculum.

Colin, on the other hand, dismissed CAL software as 'parrot fashion' software and resisted its use. This difference in viewpoint opened up a conflict in the philosophy of computer use between Colin and John: interestingly enough, the means for resolving this conflict lay in the separation of Logo from mathematics.

The new policy suited both teachers, but for different reasons. Colin was swung to an applications point of view; John demanded CAL for mathematics. So the change from LogoMathematics to Logo-IT could be traced to the fact that the school personnel, though initially keen on the innovation, were not convinced that Logo in mathematics was not simply Logo 'as an experience'. They did not share the aspirations for the software held by Claire and ourselves, and neither did the head. If they chose to see Logo as unconnected with mathematics, they correspondingly chose not to focus their students' attention on the mathematics of the Logo situations they encountered; Logo as a catalyst for webbing mathematical ideas simply did not enter the picture. On the contrary, self-fulfilling prophecy became reality: the Logo work *did* have nothing to do with the mathematics.

It is worth asking why the history of the Harston experience differed so radically from the successes of the teachers on our *Microworlds* project whom we met in Chapter 8. One difference stands out as fundamental. The *Microworlds* work existed to re-vision the mathematics curriculum: the knowledge domain of school mathematics became a subject for negotiation between most (not all) of the teachers, and with ourselves. In Harston, the reverse was true. The curriculum was a given, and success was measured by achievement on 'it'. Apparently, the Harston teachers were more prepared than the *Microworlds* teachers to view mathematics as a fragmented set of topics expressed in questions such as 'What is the sum of the angles in a triangle?'; and they were correspondingly less prepared to consider possibilities for change within the curriculum.[6] The meanings of mathematics were fixed, and any innovation had to be constructed in this light. If students were judged according to their successes and failures on a fragmented curriculum, then Logo had to be judged in the same light. Given the immutability of that curriculum, it was Logo that had to change.

2.3. *The advisory teacher*

Having decided to invest in the project, the Local Education Authority appointed an advisory teacher, Claire, to visit the school at least once or twice a week during the first term. Claire was knowledgeable about Logo and deeply committed to its use in mathematics. What turns out to be important in our story is that Claire had previously worked in Valley and had been a prime mover in the earlier LogoMathematics initiative there. Her primary-school background and her commitment to a certain way of working with Logo was clear from the start. Just what this way of working was, and how far it differed from the Harston approach, soon became evident:

> ...Well I started off visiting once a week, sometimes twice a week in that very first term, but definitely once a week, literally to be there as a support if I was needed, moral support, moral more than anything and ... I tried to give support in terms of organising the classroom in order for the Logo to happen but failed dismally...

Claire had very strong views about the teachers with whom she worked. She felt they resisted thinking about strategic organisation and resorted to 'masses' of Logo worksheets to introduce the children to as many features of the language as possible, rather than projects about mathematical ideas incorporating Logo.

However, talking about mathematical practice appeared to be a necessary but not sufficient condition for success. John was well versed in the philosophical debates and indeed espoused views not dissimilar to Claire's: yet Claire talked with despair of John, describing him as simply responding in ways he felt were politically correct, rather than making these ideas his own. She was frustrated that John had taken over Colin's worksheets (even though they may not have been appropriate), did not think about his intervention style, nor work out how to organise the children's time on the machines. His response to the innovation, in her view, was 'to keep it at a distance', to engage in discussions about the philosophy underlying the innovation in terms of generalisations which did not engage with any implications for his practice:

> I mean... the children didn't know what they were doing, they couldn't remember when they'd been on the machines, they didn't know what worksheets they were supposed to be on and they didn't know what they were supposed to be doing... He gave the impression of 'knowing everything' and in fact he 'said the right thing'... but I mean his words were very different, the difference between rhetoric and practice was enormous.

Claire's frustration with John was clearly fuelled by her practical approach – for her, *doing* was sufficient goal. A further difficulty, one which ultimately brought problems to a head, was that of integrating the Valley students with the rest. Claire felt that for many Valley students it proved to be a matter of marking time, repeating work and consolidating what they already knew while the others caught up. Claire surmised that the school thought:

> ...they had all to be pushed to the same level and so actually pushed these other kids through recursions, spirals and goodness knows what... so that

they could all do the same thing. It almost seemed like a competition with the primary school in that Harston could cover all the 3 years' Logo work in the junior school in one term!

Perhaps there was an element of competition exacerbated by Claire's prior involvement with Valley? This involvement was originally conceived as an advantage for the innovation to foster continuity of ideas and practice. But underlying many of the obstacles encountered was a clear difference between Valley and Harston; Harston's keenness to assess, to attach children to levels and to ensure 'progress', was radically at variance with Valley's philosophy. At Harston, a tick list was generated whereby coverage of certain Logo topics (e.g. spirals), meant going into the top group: such a move would, Claire felt, have been unthinkable to Valley's teachers.

By the second year of the innovation, Claire had the impression that little new was being undertaken and felt the teachers were not in a position to develop mathematical microworlds. She saw these as crucial to the success of the venture, but because of her negative feelings towards the Harston situation, she apparently could not work with the teachers on this task. She noted there was little monitoring of who was spending time on the computers, that the brighter children and the boys were taking the lion's share of the computer time, that the teachers competed over scarce hardware and that there was little discipline or sense of direction. Yet she did not feel inclined to intervene. She even gave up talking to the teachers when she went to the school:

> I worked with the children on the machines, pointed out to Colin that he never spoke to a girl, and that was about it. I stopped going to John's quite early on.

Claire was aware that the innovation was dying out, even aware that blame could not simply be layed at the teachers' door – there were new and increasing pressures on them:

> Gradually the work was fizzling out except for children who demanded access, there was no decision to abandon the innovation but it was becoming unsatisfactory and changes in examinations were taking up time and energy.

Why did this happen? Claire clearly blamed the choice of school:

> I mean it was the wrong place to put it in, had we have had the choice that's the one school we would not have put it in... Because everything's against... everything is wrong... in terms of a new innovation... the head is not a forward-thinking educator, he very much... feels himself pressured by the parents. So... he's definitely more concerned with outside image than actually what goes on in the classroom for the benefit of the children.

> There was also a feeling of lack of direction and an absence of a cohesive policy in the school – or least one shared with the staff as a whole... also there is streaming and not much computer confidence.

Claire expressed bewilderment at what she saw as the distance between theory and practice – how the Harston teachers expressed pride in the innovation in meetings about the Logo project and were heated in opposition to people who expressed reservations; yet this energy did not, in her eyes, spill over into their practice.

It would be wrong to take Claire's view at face value; she, like the Harston teachers, brought meanings to the project which were moulded by her educational experience and culture. Perhaps she could not detect energy in the school, as she was looking for outcomes which differed from Harston's. Maybe she could not assist because of her different philosophy, the competition which emerged between Valley and Harston. Perhaps her expertise was even devalued in the eyes of the secondary teachers as it was based on primary rather than secondary experience. Her lack of intervention was clearly not lack of motivation:

> I mean I think it's proved a disappointment and if I could have willed it... wished it or whatever to have worked more... I would have done because I really wanted it to, for Logo to be good enough to survive.

3. THE STUDENTS' VOICES

To obtain the students' views on the changes in the autumn term of 1988, we interviewed 6 children who had recently entered their third year at Harston (they were now aged 15 or 16), but had all been at Valley where they had been part of the original Logo project. All the boys were doing well, all had opted for computer studies, although Michael, who had been one of the stars of the Valley Logo group, was now regarded as a 'bit of a problem' and Daniel, who had been the most brilliant Logo user had become 'bored with school'. Sadly, both the girls interviewed appeared rather disillusioned with school in general and with Logo and computers in particular. Neither had opted for Computer Studies, despite all their earlier Logo successes. Jane simply said she was bored and was glad to drop computers, and Norma had been dropped from the top mathematics set and proudly announced that she had not used a computer for 2 years.

In trying to make sense of the data gleaned from the student interviews, we distinguished four categories as organising themes, which reflect both our theoretical framework and the students' viewpoint as it emerged from the interview data. The themes were: the meanings of programming; meanings of LogoMathematics; teaching and the trajectory of Logo learning; and classroom culture. We will comment on these in turn.

3.1. *Meanings of programming*

Two sub-themes can be discerned under this heading: how students' views of Logo contrasted with their views of BASIC, and how far students felt they had become competent in the Logo language (and how far they valued this competence). All the boys, but neither of the girls, had opted for computer studies and many contrasted their present experiences with BASIC with their previous Logo work. For example, Michael found computer studies hard and boring:

> ... a lot of it is just going through the computer book and finding how computers work and it's... I find ... some aspects of BASIC very, very hard, I get a bit confused with... sometimes there's loops and things... I find that difficult.

He recognised that his previous Logo experiences had helped him: 'it gave me a basic understanding of quite a lot of things ... about computing about structuring programs, about ... procedures and things and what different programs do, different types of programming things like that ...', but regretted that he had not really explored the full potential of the language. He felt he had reached a plateau in his Logo work and had only just 'started getting the hang of text [list – CH/RN] processing and things like that you know, not just graphics but other... databases and things...'.

By way of contrast, Jane felt that her Logo work had rather led her into a cul-de-sac. She constructed Logo as a language for young children; merely an introduction to programming and not a very good one at that!

> Yeah, I don't... we didn't get too far on Logo really, I think doing actual computer studies where you learn how to program... would be more important than Logo... it was like a child's version of it, if you know what I mean... I'd like to learn sort of like proper computer programming again... Yeah, 'cos Logo didn't seem to me that it was a proper adult thing... it didn't seem to me it was an adult thing that would help in business or... you know it didn't seem that it was helpful, well it was helpful in some aspects but not sort of what I would sort of consider when I was older.

Lawrence disliked computer studies for a variety of reasons: the book work, the fact that he was not quick at BASIC and the style of having to do 'dry runs' before touching the computer. He was clear about what he had learned in his Logo work in Valley, describing the 'logic' needed and its connection to investigations and projects:

> Well I learned a lot about how the computer works, the way you have to use logic rather than you know, you can't sort of force your way with the computer and you have to like... work it out first and then sort of see if that works and if that doesn't work... you know you want to find things out... and I think that's what I enjoy about lots of things.

He was sorry that they had not got further and regretted that Logo was to be restricted to the first two years of Harston:

I would have liked to get further on to Logo 'cos apparently it's quite a high level language I mean, and we only scraped the surface, you know... and I think we could have used it a lot better, you know... better to go on with it further and now we've got the Nimbuses[7] you could do that, it's just been dropped.

All these responses were echoed by Peter who admitted he had found it hard to learn BASIC, at least partly because of the programming style imposed:

Well once we've... first of all we talk about how the program will work and I find that bit quite easy ... the logical thinking. But then we have to write down the program on paper, and go through it to make sure it'll work before you're allowed anywhere near the machines. And I find that difficult. I'd prefer to be able to go on to a machine and try it out and if it doesn't work try it a different way, but we're not allowed to do that we have to write it down on paper first.

What made matters worse was that Peter had picked up that BASIC was not very useful in the outside world, so he felt quite resentful at having to learn another language when he could have continued with Logo:

...and I didn't like the idea of switching to BASIC after I'd used Logo, I wanted to stay with it 'cos you've got to a stage where there's a certain amount of things you can do and you just want to go on... developing that and then you have to sort of start from scratch again. I think it would have been good if we could have carried on with Logo, through computer studies...

This irritation at not having fully exploited all the possibilities of Logo as a programming language was echoed by Daniel, who had become an expert Logo programmer a few years earlier, largely by his own efforts. He now realised how much more he could have learned and was a little upset by this:

I've looked through the manual reasonably recently and I've found that there are quite a few other commands which you can use.

He was left frustrated that he had not been guided to all the nooks and crannies he could have explored.

3.2. Meanings of LogoMathematics

After our analysis of the student interviews, it became clear that despite spending a couple of years doing Logo in mathematics classes in Harston, most of the students expressed few if any connections between the two domains. The meanings they brought to mathematics were formed at Harston;

those for Logo were derived from Valley. It turned out that these two sets of meanings were very difficult to reconcile.

Michael regarded mathematics as hard. Mathematics was about working out problems involving well-specified content and linguistic structures, not unlike programming: '...it's like doing if-then conditional statements in BASIC'. But Logo was different: it was straightforward, fun and investigative. Similarly, mathematics for Norma had set routines and right answers so Logo was more like science, clearly distinguishable from mathematics:

> No, maths is more, you know you're told what to do and then you have to learn it.

Lawrence's perspective was strongly linked to affective concerns and he explicitly contrasted his enjoyment of Logo with his gloom in mathematics, recounting gleefully:

> I got to stay in the break (to do Logo) and you know, get off doing maths and things to go on the computer.

However despite rejecting connections between Logo and mathematics, Lawrence did admit that he still used Logo metaphors to help him in his mathematics:

> In grid references... 'cos it was all move and draws... and I can work them out... 'cos of my experience in Logo... I think, anyway.

When thinking about what he had learned with Logo in Valley, Peter stressed building up his confidence with the machine and failed to mention anything about mathematics:

> I think the main thing was I felt comfortable with a computer... it wasn't you know, sort of keeping away from the computer you felt comfortable using it, it wasn't a machine that was difficult to understand.

For Peter and Jane, mathematics was defined by its topics and its content: Logo simply had no connection to it. What was rather strange was that during their first year in Harston these students had visited the University to undertake some experimental work which *did* have a very explicit mathematical agenda. Yet it was no longer accessible to them, it was forgotten – perhaps since their attitudes and perception were already firmly in place.

Daniel was the exception. He found it simple to make connections between mathematics and Logo since he considered mathematics as a tool to be exploited within his projects. Logo for Daniel was project-oriented; so mathematics emerged rather naturally during the course of the work:

> We didn't start off thinking we were going to do a scaling graph program but then as soon as we thought of it 'oh yes we did that in maths, we can use that'.

He contrasted the ease of writing little tools for his projects in Logo with his frustration with BASIC (particularly with the graphics side) which he complained took considerable time, was unnecessarily complicated and produced very mediocre results.

3.3. *Teaching and the trajectory of Logo learning*

A major issue for all the students was *progression* in their Logo work. They all reported feelings of marking time at Harston while the children with less experience from other primary schools 'caught up'. Michael, for example, expressed dissatisfaction at his progress in Logo but was sophisticated enough to see why this might have been:

> I think really... we didn't progress as much as we would have done if it had... if everybody had been at the same standard, but obviously Mr Smith had to teach people who were just learning.

Daniel made a similar point:

> I began to feel that it was getting a little limited, and I wanted to do more things with words and... use [list] processing and stuff and it seemed to be that I typed some of the commands but there was nobody to teach how to use them and it was getting, it seemed to be getting a bit restrictive... after about one year of it I felt that there wasn't much more to do in it, you could write more programs but there weren't any more commands to learn.

Again, he mentioned the problem of pacing in Harston and the need to keep everyone together:

> I used... it used to annoy me when we all had to do the same thing in maths, and I didn't enjoy that... I wanted to go and lock myself in a room with the computer and just type what I want to do and what I want do to on the computer. And Mr Smith would make us do what the rest of the class were doing and that's the only time it used to annoy me when I had to go back over things I'd already done.

Jane also had a very clear recollection of the Logo work in Harston as repetitive of her previous activity in Valley:

> It was just doing the stuff that we'd done at Valley really... we were learning more at Valley... I mean... we were learning how to program ... at Harston we just stopped learning and we were just carrying on doing the same thing, drawing pictures, but I'd learnt how to do it... but we didn't learn any new things.

Both Jane and Norma attributed this to the same cause. As Norma put it:

> ... because there was only a few of us that had come from Valley, and everyone else hadn't done anything with computers, so we had to wait for them to catch up, you know.

Peter enjoyed his first year at Harston when all the group from his primary school had kept together in one of the mathematics sets, but had noted how the momentum had slowed down in the second year. Strangely enough he seemed very knowledgeable about the policy discussions underway in the school:

> It was just us working on it, it was good. I enjoyed that. And then it started to break up and the course came to an end. It was a bit of a shame 'cos you know all that work ... I think it was about at the end of the 2nd year, half way through the 2nd year. We were... they were trying to decide whether in the next year there would be an actual computer studies lesson or whether computing would be done in maths, and I think they were trying to decide between if it was done in maths it would have been Logo I think... or else the computer studies lesson which was BASIC ... If it would have been done in maths we wouldn't have got so much time on computer 'cos we only had 5 lessons a week in maths and we've got to work through the course in maths so I don't think we'd have got as much computing. Whereas computer studies we've got 4 lessons a week, actually in the computer room doing computer work. I think it would have been nice if we could have used Logo ... in Computer Studies, I don't know the practicalities of it but ... it would have been nice to continue the course.

3.4. *Classroom culture*

When we asked students to recall a best time when learning Logo, all of them recalled a time in Valley. As we will see, these were recalled in exceptional detail, even though they had left Valley more than 3 years previously. Why was this? What were the main characteristics of this experience and how did it contrast with their more recent Logo work in Harston? A surprisingly consistent pattern emerges of how the students saw the Valley Logo activity. They emphasised *enjoyment* and *challenge*, the *hard work* they had put in and the *fun* of the usually collaborative enterprise. Michael for example remembered with pleasure his sprite and quiz programs:

> I think... when we first started using the Atari with Sprites... and me and Daniel did that program with a man throwing a javelin and running and things that suddenly opened up a whole new... that was really good fun.

As with the other children, Daniel's most exciting time was way back in the primary school, where he recalled the challenge and effort put into his clock project:

> My clock definitely ... no two ways about it ... it was just a picture of a clock and it kept time and it just moved the hands ... it looked really boring and then you'd look at it and you'd see it start moving, but that took me a long time. I thought that was really complicated ... yeah, but you have to move the hour hand a little bit every time you move the minute hand.

The interview with Norma was rather difficult as she had by this time developed a negative attitude to school. The only time she 'came alive' was when she recalled her Logo project in Valley, remembering her enjoyment: 'writing my own programs and actually seeing it on the screen'. Her favourite time had been when she had designed a quiz and asked all the class to join in to try to answer her challenges. There was a feeling of competition and accomplishment both in the construction of the quiz and its operation with all her peers. Peter's best Logo time was, once again, a Valley project:

> The javelin program and the superman program. I just liked it, it was very creative ... and there's a lot of my idea which it wasn't before it's normally sort of like me and Dan, but this time I had lots of good ideas for it, like the javelin going up into the air and then rather than getting the javelin to move or anything to give it some feel we made trees and houses rush past ... The superman ... that was originally Daniel's idea ... he'd drawn all these skyscrapers, and I said that looks great, that looks like something out of superman. He said "yeah good idea" and then I worked on the sprite... we were trying to work out the tune for ages... and I couldn't remember how the tune properly went ... it took a while but I ... we had to use something like 6 sprites for 2 sprite lengths, you know 'cos of all the different colours and the cape and everything.

Lawrence described his best work as his space invaders program stressing all the hours he had put in and the fun he had:

> ... we did it with the sprites colliding and everything like that, and it was good fun, I really enjoyed it.

Jane too recalled her 'netball project' at Valley:

> ...... I remember the ones I did at Valley like the netball... We did a program where it was pupils and they were throwing a ball into a net... I think it was helpful then, because we were tackling things that we ... that were quite ... that we didn't know how to do and we were ... solved them and we felt, you know, we'd done something and we were learning more and that. And at Valley I think we learned more than I did here ...

For Jane, another distinguishing feature of the Valley Logo work became apparent – her *ownership* of the project:

> Well you know, we learnt ... we tackled more problems and we thought of a thing we wanted to do ... we got helped along to do it.

She contrasted these projects with those in Harston where ideas were imposed:

> ... but when I was sort of in 2nd year (here in Harston) we were just doing like ... we did... a Christmas card and what we just had to do was draw a picture of a Christmas card and put it into a program and it wasn't ... They told us we had to do a Christmas card.

Norma also contrasted the fun in learning, clearly linked to an open-ended investigative approach with the more problem (answer)-oriented strategies in Harston:

> I think at Valley where we started off we were younger and we were doing turtle graphics it seemed to be more emphasised on fun ... it's like fun through learning you know, we really enjoyed it. We were always fighting to get on to the computers ... whereas when we got to Harston I found that ... types of programming changed and it was an emphasis on solving problems ... they wanted you to get an answer to something.

Peter spontaneously compared the cultures in Harston and Valley. He remarked in passing:

> It's strange, I seem to remember all the projects from Valley but none from here...

When we asked him to speculate why this might be, he pointed to the culture in Valley, the significance accorded to the Logo work:

> I don't know why that is; maybe it was the way it was taught ... because it was ... at Valley it was a very sort of ... high on the agenda sort of thing, and the computer was in the room with us all the time, whereas now we only go to the computer room occasionally.

Other students talked with similar enthusiasm about the culture of the classroom in Valley. Michael tried to express why it had been so positive:

> Yeah, because everybody in the class did it and it was kind of much more... the teachers were much more interested in it, and we were doing it as a whole.

Although there was only one computer in the classroom, this had not been an obstacle, as the ideas from the Logo projects had pervaded all the work:

> No, I don't mean at the same time but because we all ... everybody in the class got to use it, then quite a lot of the time our teacher was talking about computing to the class.

Lawrence similarly observed that his teacher in Valley accorded great significance to the Logo work, an awareness which had been sharpened not only by the practice of the teacher but also by all the paraphernalia surrounding a research project. When Lawrence talked about his Logo work in Valley he said with huge conviction how much he had loved Logo – not only for its challenge and way of learning but also because of all the attention he had received as part of a project from visitors, films were made etc. Peter also alluded to this outside attention and he admitted that he had enjoyed all the limelight:

> Yeah ... the chosen few... we were chosen to have this project and we had ... people coming in and ... people were writing a book and they were coming in and speaking to us, and I enjoyed going up to the University in London, I really enjoyed that.

Little was mentioned of the culture in Harston except by Jane who commented that she could remember nothing she had done at Harston: her views of Logo in Harston had been strongly influenced by a feeling that Logo was not regarded as important there – either for itself or in connection with mathematics:

> So it wasn't that important sort of ... it was just like you know a little exercise for us all to do, so we'd all had experience on the computer I think.

4. MEANINGS IN CONFLICT

One way of describing the Harston story is based on a simple conflictual model. The Logo work was never seen as part of mathematics or even as mathematical; rather it was seen as taking time away from mathematics teaching. Its separation into IT satisfied the push to programming instigated by the head of computing, and conceded to the mathematics department's reluctance to embrace a pluralistic view of the mathematical content they felt they had to prioritise. Although part of a LogoMathematics team, the teachers found themselves in the middle of an innovation where the multiple meanings imported from diverse sources simply could not be resolved. The LEA and University staff, the head and the teachers all faced a continual series of conflicting tensions: the responsibilities for pupil learning, covering a given curriculum, accountability to parents. Support was available but it did not offer what was required – the University ideas may have been seen as 'out of touch' and the adviser as 'too primary'.

Yet while this simple description certainly fits the circumstances up to a point, a simple conflictual model does not explain the transformation which occurred. For the initiative did not die. Rather LogoMathematics was reborn as Logo-IT – a shift which suited the inclinations and expertise of the teachers as well resonating with outside policy and equity issues. How can we explain this transformation?

At the heart of the shift lies a struggle for meanings derived from different cultures, and re-forming within the culture of the school. From technological cultures, we saw a set of meanings for what it means to 'program', what 'IT' is for, how Logo was a 'spent force' (we have no evidence that the headteacher was aware of the post-Pea-and-Kurland culture we discuss in Chapter 7, but his remark is suggestive of the extent to which the ideas had spread through the system). There was a set of epistemological tensions concerning the nature of mathematics, the role of curriculum, why students should learn mathematics. From the students' point of view, the change in culture between Valley and Harston led to confusion and tension with respect to diverse issues such as styles of programming, attitudes to mathematics and Logo: (for example, mathematics was hard, right or wrong, routinised and boring; Logo was experimental, challenging and enjoyable; in mathematics problems were given, in Logo 'you could pursue your own projects'). From the teachers' point of view, there were tensions between epistemologies (mathematics *versus* computing), and pedagogies (delivering the curriculum or changing it). And there were broader, social and political concerns of competition between schools, parental pressures and local accountability which intersected these tensions and dilemmas in unexpected and sometimes surprising ways.

We have met the proponents of this struggle for meanings: the advisory teacher, the headteacher, the parents, staff and students of the school. They remind us that there is an interplay between the cognitive and the cultural, the ways in which webs of meanings constructed by individuals are derived from and help to reconstruct meanings within cultures.

We have seen a snapshot of the concerns, aspirations, and constraints operating in Harston. We saw, for example, how the egalitarian aspirations of Logo use as an investigative tool in mathematics, became transformed in the early days in Harston to Logo as a specialised tool for a bright group of students, particularly boys, who managed to push forward on their own and dominate the scarce resources. The school's concern with this monopoly was again made apparent in the separation of mathematics from Logo. Although we have no data to illustrate it, we can surmise that these tensions were not new; on the contrary, it is likely that the introduction of the LogoMathematics initiative did little more than evoke a particular set of tensions and dilemmas that the school already faced.

As these tensions surfaced, the particularities of the innovation – its goals, approaches, rationales and organisational forms – were moulded and reconstructed in the course of countless struggles and dilemmas. Any thought that we could study the trajectory of one thing (the innovation) into another (the school) must be dismissed: the situation is much more complex, involving a dialectical interaction between the two, a reconsideration of how the innovation shapes and is shaped by its entry into the school setting.

There is another dialectic at work. Thus far, it might be thought that we could envisage each individual or group bringing their own set of arbitrarily-defined meanings to the situation. If this were true more generally, it is hard to see how common discourses could ever emerge within a given culture. But the meanings brought to the Harston setting, and to communities more generally, are not arbitrary: they are derived at the intersection of the ideas of the individual and those of the setting of which they form apart. People construct their own webs of meaning, but only in terms of the webs they already have to hand.

If we view the innovatory process in this way, we can begin to see how the transformation we have described took place. Resolution was possible – but at a price: Logo as a replacement for BASIC, mathematics as an immutable system, new pedagogies embraced and rejected. The school culture strove to make sense of these meanings, much as an individual strives to fit the events of their intellectual and social experience into a coherent, meaningful pattern. Schools, like individuals, are seldom autistic: they have to reconcile the meanings they encounter. And if they cannot do so in the terms laid down for them (e.g. by researchers or advisory teachers), they have little alternative but to reconcile them in some other way.

This way of looking at the events at Harston resonates with the case studies developed in Chapters 7 and 8. In Chapter 7, we showed how the educational-political cultures transformed the LogoMathematical idea, and how elements of the research community constructed a particular set of meanings for Logo-based research. In Chapter 8, we illustrated ways in which teachers' personal meanings for their activities were derived from the cultures of which they were part, and in turn, how they acted back upon these cultures. Now we have seen how Harston school approached the LogoMathematical idea with an unstable and emergent set of meanings and how, in the struggle for stability, they fundamentally reshaped the idea itself.

Whereas the conflictual model leads to inevitably pessimistic conclusions, our alternative leaves open some grounds for cautious optimism. For it allows us to think of meanings being created within a setting, rather than simply brought into it: if the latter were the case, we might be convinced that some changes are simply not possible due to the inherent conflict between inno-vation and setting. But meanings are *created* not simply *received*. Change is possible. In other words, it is possible to outline a program for the construc-tion of new, more learnable mathematics; and it is to this challenge that we now turn.

NOTES

1. The *Chiltern Logo Project* was funded by the Microelectronics Education Pro-gramme. RN was the principal investigator.
2. The names of the schools and the participants have been changed.
3. The designations 'first year' and 'second year', apart from being confusing to non-UK readers, are now obsolete in the UK. Valley school was a 'middle school'; the children arrived at Harston aged between 12 and 13.
4. Information Technology awareness consisted of the usual collection of 'basic skills' such as keyboarding, coping with operating systems, and handling disks. The Computer Studies Option was still on offer although in line with national trends its future was under review.
5. We hope the reader will appreciate that we are poking fun at ourselves here, not at Colin.

6. We accept the possibility that changes we observed in the *Microworlds* teachers may not have been robust, as we were only able to make minimal observations in their schools. We did follow-up the teachers after one year, and interviewed them some three years after the course, when we found they espoused much the same views as they had earlier.
7. A British-manufactured PC clone. It is interesting that Lawrence, like his teacher, sees the advent of new hardware as potentially solving his problem.

CHAPTER 10

RE-VISIONING MATHEMATICAL MEANINGS

revise (-z), v.t., & n. **1**. Read or look over or re-examine or reconsider & amend faults in (literary matter, printers' proofs, law, constitution, etc.; $R\sim ed$ *Version*, abbr. R.V., revision made 1870–84 of Authorized or 1611 Version of Bible); hence or cogn. $\sim$'ABLE, $\sim$'ORY, (-z-), aa., $\sim$'AL(2) (-z-), **revis**ION (-zhn), nn., **revision**AL (-zho-) a., $\sim$'ER1 (-z-) n. (esp. in pl. of authors of R.V.). 2. Revision, $\sim$ing, (rare); $\sim$ed form (rare); (Print.) proof sheet embodying corrections made in earlier proof. [n. f. vb, f. F *re(viser* look at f. L *videre vis*- see), RE- 8]. (*Concise Oxford Dictionary*)

1. REVIEWING THE SCENE

By exploiting the computer as a window on the multiple ways mathematical meanings are constructed, we have drawn together the different strands of our work in an attempt to formulate a unified theory of mathematical meanings. We have proposed *webbing* as the pivot of this theory. The idea of webbing emerged during our struggle to reconcile the individual's role in the construction of mathematical meaning, with the recognition of the part played by epistemological, social and cultural forces. Webbing tries to convey the existence of connections built into the structure of any environment, signposts which assist in navigation: the signposts followed and the connections reconstructed for support stand in a dialectical relationship with the actors in the system.

Let us briefly review the main elements of our story. It began with a critique of global hierarchies in mathematical meaning-making. The perception of mathematical learning as the acquisition of context-independent knowledge within a hierarchical framework is, we argued, an inevitable outcome of an exclusive focus on epistemological and psychological facets of webbing which pays no regard to social and cultural sites. Moreover we regard it as hardly accidental that these hierarchies mirror sequences in the school curriculum, since these too have the same foci: the individual and given mathematical knowledge.

The problematisation of hierarchies raised the question of the meaning of abstraction. We wanted to retain abstraction as central to mathematics; yet if we rejected abstraction as ascension and loss of meaning, how else could it be characterised? Following Wilensky and others, we have proposed the notion of abstraction as a process of connection to new objects, a process that develops in activity: abstracting within a domain rather than away from it. This suggests that there is more than one kind of abstraction – being abstract cannot be defined *a priori*. Our notion of situated abstraction encompasses new ways to abstract, and the creation of new meanings. We can, therefore, extend the scope of what it means to do mathematics without sacrificing those characteristics which give mathematics its power. We can glimpse new knowledge structures, new ways of making meaning: we can re-vision mathematics.

We have thought of learning to participate in mathematical activity as learning to articulate structure and relationships, and we have sought to describe conceptual worlds that possess this expressive power. In these domains, abstraction is situated in a system of signs and practices, which, as in official mathematics, play a critical part in structuring which relationships are expressed, and how.

Webbing and abstracting are complementary. Situated abstraction describes how learners construct mathematical ideas by breathing life into the web using the tools at hand, a process which, in turn, shapes the ideas. Tools are not passive: in a microworld, for example, the designer's intentions are constituted in the software tools. These tools wrap up some of the mathematical ontology of the environment and form part of the web of ideas and actions embedded in it. Yet it is students who shape these ideas. We have given examples of how students can act within the webbing of the medium, sketch a solution strategy, flag its critical features or capture its patterns. A microworld comprises tools to construct objects. But these tools are themselves objects which encapsulate relationships. This process/object duality is at the root of mathematical activity, and we have shown how interaction in microworlds can transform this duality into a dialectic. Webbing involves stepping on to a structure which has already been erected — the software webs the actions – but it leaves the critical step of seeing the general in the particular, the theoretical object in its construction, in the hands of the learner.

Considering abstraction as situated, immediately raises the question of connections to official mathematics. We may see the generality beyond the tool; but why should users build connections outside the microworld? To generalise from one setting to another, to become aware explicitly of the relationships wrapped into the setting, means to notice precisely those elements of the computational web that are interacting with one's current state of understanding. The point is that the web is not local – it is hugely global and interconnecting; although expressions are constructed locally, the metaphor of webbing points beyond any particular situation.

Clearly there is a role for the teacher in drawing attention to appropriate connections in the web. But it is by no means unproblematic to achieve the delicate balance amongst connections needed to develop mathematically; when the connections to the teacher's goal are dominant, students may simply imitate a procedure or try to guess what the teacher wants; when they are not made at all, any mathematical agenda might be bypassed. We cannot ignore the social dimension of mathematical meaning; how the presence of others mediates the growth of an individual's web of meanings by catalysing new knowledge-connections between individuals.

The computer is not only a set of tools which defines allowable actions and their expression; it plays a role in communicating the actions, sharing and re-negotiating mathematical expression and facilitating the (co-)construction of mathematical meanings. We have shown how collaboration is intertwined with the activity structures of the setting: people collaborate to achieve meaning in ways which relate in a surprisingly fine-grained way to the knowledge structures and types of task with which they work. Rethinking collaborative work pointed us again in the direction of epistemological revisions.

We used the computer to open new windows onto more general, systemic appreciation of the ways mathematical meanings are constructed in the broader educational culture, at the ways in which the educational system comes to construct meanings for school mathematics. In the last three chapters, we studied the ways in which Logo had been inserted into educational cultures, and how the cultural phenomenon of Logo and the education system mutually transformed each other. We looked at the ways in which some corners of the educational system – the research community, the curriculum, the school – constructed meanings for Logo and with them the mathematical ideas it was intended to illuminate for learners. Here too, the epistemological dimension emerged as paramount; the epistemology of Logo and that of school mathematics interacted to shape the meanings of each.

We can think of this meaning-making process as an extension of the webbing idea to the cultural domain. Connections are formed between ideas from disparate cultures, and these come together to create meanings at their intersection. One of these intersection points is the teaching community. How teachers interact with any innovation depends on a multitude of influences, not least the constraints of the classroom. These constraints are reconstructed by teachers as they take on board those aspects of mathematical meaning-construction that fit with their expectations and beliefs – the crucial point is that these too are subject to change. Through the window of the computer, we saw how meanings for the computer and for mathematics were created by teachers, and how these shaped the relationship between the computational media and the mathematical knowledge they were responsible for teaching. We saw how change was revisioned; but its trajectory was variable as different teachers forged some connections while ignoring others. Once again, the epistemological dimension came through as central. It became clear that where teachers revisioned mathematics, they breathed new life into their professional practice.

Finally, we offered a case study of the introduction of Logo into a single school. At its heart lay a struggle for meanings in the course of which the innovation shaped and was shaped by its entry into the school. The meanings created were neither arbitrary nor predetermined, but derived at the intersection of the actors' beliefs and behaviours – teachers, students and outside agencies – and the setting of which they formed a part. There was a strong epistemological thread running through the case study, framing the ways that the knowledge domains of Logo and mathematics were conceptualised.

One message is clear. A key explanatory variable in tracing meaning-construction across all our settings is epistemological. Our concerns have continually cycled around the kinds of knowledge which can be gained from practices, how knowledge in different domains is connected, and the relationships of knowledge to the media in which it is appropriated and expressed. More fundamentally, we have seen how mathematical knowledge was constructed in action and reflection, not as a unitary and inert thing-to-be-defined. There are underlying structures of knowledge, 'epistemological forces' as diSessa (1995) puts it, which are mediated by the lived-in-cultures of individuals and groups.

Our notions of webbing and situated abstraction fit into this picture. They explain how it is possible for individuals to participate in a cultural practice, and at the same time build more generalised systems of knowledge which connect with the acceptable norms of the knowledge structure in the prevailing culture. Our preference for viewing this process through a computational window has provided a convenient arena for understanding how the resources of the media structure the interaction, as well as the process of knowledge building. For those who are not yet enculturated into the complex and finely-tuned discourse of official mathematical language, such domains – especially if they are carefully designed from a pedagogic stance – offer rich and complex settings which can act as bridges to, rather than barriers between, traditional kinds of mathematical expression.

2. EPISTEMOLOGICAL REVISIONS: SIX EXAMPLES

We now describe a selection of epistemological revisions of mathematical activity, all of which derive from interaction with computational media. The first two examples focus on the relationship of tools to mathematical knowledge; the third and fourth shift the focus to school mathematics. These four revisions point to new connections within mathematics, new ways to express relationships and structures by means of the available tools. In the fifth example, we illustrate the possibility of connection between abstractions rooted within the computational medium and those of official mathematics. In the sixth and last example, we consider an epistemological revision in the context of mathematics teaching within a particular cultural practice.

2.1. *Small changes in tools afford large changes in meanings*

Our first example concerns a Logo microworld for exploring non-euclidean geometry, based on the doctoral dissertation of Ian Stevenson, a student and now a colleague of ours at the Institute (Stevenson, 1995). We choose it for several reasons, not least because it connects with an area of mathematics which is, in comparison with some others in this book, 'advanced'.

Non-euclidean geometry is an interesting area of study since superficially it would seem difficult to mobilise intuitions to make sense of it. The visualisation of abstract mathematical structures such as non-euclidean geometries is difficult, though – as Stevenson shows – not impossible. There are standard models for elliptic and hyperbolic geometry which have existed for over a century, obtained by projecting surfaces of constant positive and negative curvature onto the euclidean plane. The problem is to find ways in which learners can develop a sense of familiarity with the new metrics associated with them.

Stevenson carefully describes the process of iterative design and construction of a computational world in Object Logo[1] to achieve this objective. He analyses three cycles of development over a two-year period, during which technical, pedagogical and cognitive issues were considered and reviewed, and he shows how students achieved fluency of expression with the microworld at the same time as they restructured their understandings of the geometry of curved surfaces and their projected images.

The basic idea is that the learner can vary the turtle's behaviour to correspond to the choice of an underlying model of the relevant geometry. The analogy which proved most helpful was that of a 'hot-plate' universe for elliptic geometry, in which the turtle's measuring rod expands as it moves further and further from the centre of the screen model. Similarly, in a 'cool-plate' universe for hyperbolic geometry, the rod contracts as it moves radially away from the centre of the screen.

We will focus on one insight derived from one of the cycles of the design process. It concerns one of the fundamental perceptual characteristics of the model employed – the turtle's measuring instrument varied as a function of its position. This implies that moving forward one turtle step produces different results depending on the position of the turtle (and, of course, the model it was moving in). In the hot-plate universe, for example, the turtle would take longer and longer steps as it moved nearer to the edge.

Now, as we might predict, this kind of behaviour is far from clear to most learners. For example, as they tried to draw a triangle, students could *see* that the size of the steps varied with position, but were unable to *read* the behaviour as part of a consistent pattern. As it turns out, one simple change made a considerable difference: the addition of a 'dashed' turtle track. The length of the dashes was made to correspond to the size of the measuring rod at a given position, so that the variation in distance measure was evident on the screen.

The fundamental insight is that the turtle's 'temperature' could be read from the screen; there was something to see, to experience, to manipulate. And this one new element provided just the right level of support for understanding something more about the underlying surface (one student, Sean, said: *You can tell that something's happening on the surface there...*), and the 'perspective' of the drawings made by the turtle (another student, Paul stated: *You do get an impression of some kind of perspective coming closer or going further away as you're taking the steps*).

The simple addition of the dashed turtle path enabled the students to connect the turtle's behaviour with familiar ideas such as perspective and turtle steps. Both these ideas were familiar in other contexts, though neither of them (so far as we know) had been seen as part of a formal system. Moreover, the dashed lines offered a resource for situated abstraction: instead of trying to make global sense of the situation, the students could make statements like Sean's: *It's the curve ... it's because of this step business, because we have to walk further to get the same distance. .. It's having to walk more to keep up.* The tool in the microworld environment afforded just the right resource for making a connection between spatial intuitions and the formalism of the non-euclidean geometry.

2.2. *Programming to reconnect with mathematics*

Paul Clifford is a student and colleague working with us at the Institute. As a mathematics graduate he must be considered a relatively sophisticated mathematician but, as he says, he has had 15 years to forget it all! As a school teacher he has taught with Logo and tinkered as a programmer. Paul became particularly captivated by one of our courses: 'A Computer-Based Approach to Number', which he describes as 'combining my interests in maths and Logo as well as mathematics education'. We will let Paul speak for himself:

> In the second session we considered sequences, series and the use of recursion. As a parting challenge the group were asked to look at what happens to $\sum \frac{1}{n}$ as the series is summed over more and more terms. I knew the series was divergent and even remembered how to prove this using the following trick:
>
> $$1 = 1, \frac{1}{2} = \frac{1}{2}, \frac{1}{3} + \frac{1}{4} > \frac{1}{2}, \frac{1}{5} + \frac{1}{6} + \frac{1}{7} + \frac{1}{8} > \frac{1}{2} \text{ etc.}$$
>
> So exponentially bigger collections of terms are clearly shown to exceed $\frac{1}{2}$ and the total can be made arbitrarily large. Using this idea I wrote Logo procedures that counted the first 2 reciprocals, then the next 4, the next 8, the next 16 etc. and also allowed for a simple graphical output to show visually how much the sub-totals exceeded a half (see Figure

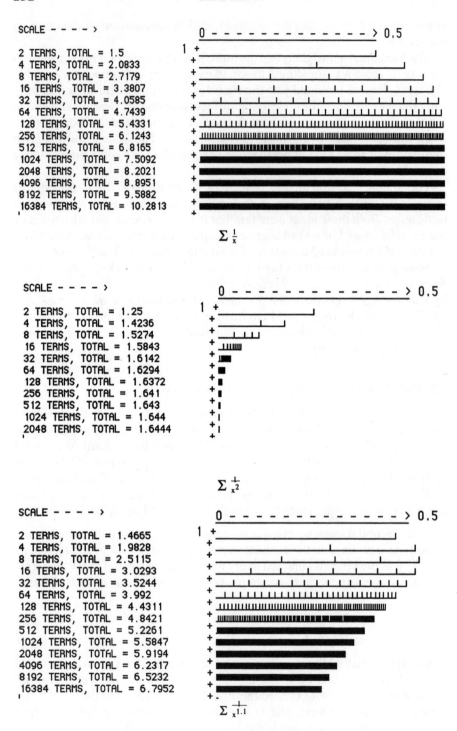

Figure 10.1. Clifford's graphical representation of his sequences.

10.1). While not telling me anything I did not already know (indeed the output does not *prove* that the partial sums do not decrease sufficiently for convergence later on), I found these results powerful visual reinforcement (in-sight). Furthermore the Logo procedures were easily extended to sum the reciprocals of any power by providing a tool to investigate the pivotal role $\frac{1}{x}$ had in this first case.

Clifford noted that it was ironic that in a course which to some extent was designed to move beyond turtle graphics, he continually used the graphics to visualise the ideas he was exploring. The key, however, was not the graphical output, but the process of construction. As Clifford put it:

The idea to look at alternative powers only came because I translated the grouping trick into a Logo procedure which could then be altered easily. The power of Logo was cumulative as slight modification brought considerable gain.

Furthermore the results suggested new questions. The graphical output seems to show that not only are the subtotals greater than 0.5, but that they tended to some constant value. The time at the computer encouraged Paul to delve into long-discarded University text books where he discovered that:

$\sum \frac{1}{x} \rightarrow \log_e n + 0.5772$ for large n $\qquad$ (0.5772 = Euler's constant.)

So for large enough n:
$\frac{1}{(n+1)} + \frac{1}{(n+2)} + \ldots\ldots + \frac{1}{(2n)} \rightarrow (\log_e 2n + 0.5772) - (\log_e n + 0.5772)$
$= \log_e 2n - \log_e n$
$= \log_e 2$
$= 0.6931$

This is the difference between the last two partial sums in the diagram. It is interesting to note that n does not need to be a power of 2 and if we marked the position of the 3500th and 7000th term, for example, they would lie one above the other in their respective rows (as they too are $\log_e 2$ apart.)

So, in exploring via programming, Paul had reconnected with emotions and ideas which he had almost forgotten: in his words, 'it has rekindled a *buzz* from mathematics I had enjoyed at school but which had waned as a graduate'. And at the same time, he connected pieces of knowledge that he had not realised were connected:

I found bits of my investigations connecting unexpectedly. Later on, while looking at prime numbers I spotted a proof that there were infinitely many primes as a result of the divergence of $\sum \frac{1}{x}$. *I do not believe that I would have seen this without my close and personal relationship with $\sum \frac{1}{n}$. It was a close friend of recent acquaintance!*

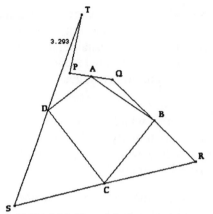

Figure 10.2. The original construction.

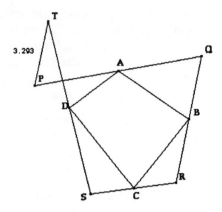

Figure 10.3. P has moved, but PT is
the same length, and seems to be
the same direction.

2.3. *New geometry with new tools*

Some months ago, we were working with Cabri on the standard geometrical
theorem that joining the mid-points of any quadrilateral will produce a parallelogram. Our first attempt was simply to replicate the traditional approach –
nice but uninspiring. We then decided to investigate how using Cabri might
help us come up with new insights into this theorem – in other words, we
started to experiment. We constructed a general quadrilateral ABCD and then
a set of line segments, PQ, QR, RS and ST which originated at a general point
P, but were then constructed so that the vertices of the original quadrilateral
were the midpoints of each line segment as illustrated in Figures 10.2 and
10.3. As we moved P around we noticed an invariant: the vector PT was fixed
(in the figures, the length happens to be 3.293).

  Immediately many questions came to mind. Why is this vector fixed? On
what does it depend? More crucially, noticing this invariance completely
changed our perspective on the old theorem. We saw that it might be a
particular case of a family of theorems. We now had a conjecture that every
quadrilateral had associated with it a fixed vector PT which happens to be zero
when the original quadrilateral was a parallelogram. Deforming PQRS with
the intention of reducing PT to zero (as illustrated in Figure 10.4) suggested
a relationship of PT to AB and CD.

  Our invariant vector seemed to be a measure of how far our quadrilateral
failed to be a parallelogram. Was PT unique in this respect? Was there more
than one vector associated with a given quadrilateral? In any case, did the
invariant vector(s) *determine* the quadrilateral? What was the situation for
pentagons – perhaps every polygon has an invariant vector? More than one?

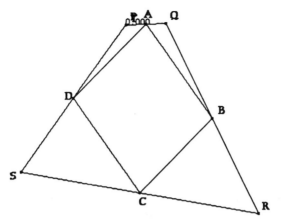

Figure 10.4. The basic points are moved so that P and T coincide. In this case, ABCD appears to becomes a parallelogram.

How many? And what about the special case where one of the sides, AD say, vanished – what was the situation for a triangle? We leave these and similar questions as exercises for the reader: suffice to say that there may well be new connections with some of the triangle theorems learned at school – theorems which had nothing to do with parallelograms!

There are still other questions of a pedagogical kind. Does this investigation throw light on how to prove the original theorem or its converse? Does it provide insight into the properties of the geometrical figures involved by presenting them in a new light? If we are to shape software with the aim of revisioning mathematics, there must be an initial stage of resistance to tried and tested approaches, followed by experimentation and questioning. Without resistance and play, software use will almost inevitably become trivialised: used to reproduce rather than revision. We can already see this happening in the U.K. In the trajectory of dynamic geometry into schools, we see little geometry and few if any powerful new ways to approach geometry. Instead we see fragmentation, traditional geometry exercises translated onto the screen. Worse still, we see the invariants of geometry located not by construction and reflection on theoretical entities but rather in the patterns of numbers generated by measurement data. It seems we are about to repeat our mistakes. In the same way that number investigations are not linked with algebraic proofs, school mathematics is poised to incorporate powerful dynamic geometry tools in order merely to spot patterns and generate cases.

But our example suggests a radical possibility too. It illustrates a way to reconnect geometry with itself, a way to model by enriching relationships within shape and space, rather than denuding them of structure. The novelty of the exploration lay in resisting the temptation merely to shape the medium to the practice, and instead allow new mathematical possibilities to flow from their dialectical interaction.

2.4. *Revisioning number patterns*[2]

A critical weakness of school mathematics lies in the gap between action and expression. We know, for example, that algebraic formulation is usually a meaningless extra in problem solving that neither illuminates the problem nor provides a means for validating its solution. Our theory of meanings points to an explanation: it is a result of disconnections between actions, the output of actions, and their description. When the web of meanings is fractured, it is inevitable that students will tend to focus on only one set of meanings: in this case, the numeric attributes of the output.

One such activity typical of the U.K. curriculum is illustrated in Figure 10.5.

Construct matchstick squares, using an appropriate number of matchsticks, to make 1, 2, 3, 4, ... squares, and generalise the pattern.

Figure 10.5. An example from the U.K. National Curriculum (Department of Education and Science, 1991, p. 10).

Our difficulty is not with the activity itself. Everything depends on the ways in which connections are built. Actions on real matches necessitate no push towards generality, no need for mathematical expression: the means of expressing actions is separate from the activity itself. It is true that in the formulation of the curriculum such (algebraic) expression is the goal: but it is grafted on to the activity itself. The matchsticks pose as a kind of visual algebra – but in practice they are nothing of the sort. The idea that playing with the matches and thinking about the visual pattern will lead to algebra is too simplistic.

We designed a microworld, *Mathsticks*, as an environment to engage with these types of activity, written in Microworlds Project Builder (MPB)[3] a new programming-based construction environment. We built icons (see Figure 10.6) representing the components of the mathematical sequences (matches, dots and tiles) which could be reproduced by the screen turtle either through direct manipulation (clicking on the appropriate icon) or under program control (typing the name of the icon in the command centre).

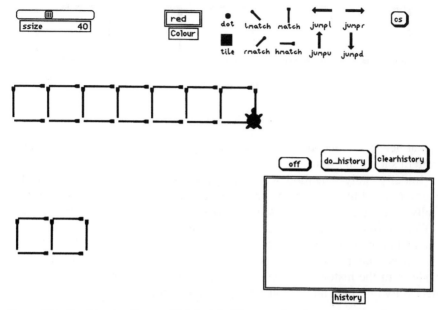

Figure 10.6. The Opening Screen of *Mathsticks*. The second and seventh terms of the (visual) sequence are shown.

We also used a history text-box to record the symbolic equivalents of actions made under direct manipulation. By clicking the on/off button (which toggles on a mouse click) students could decide whether or not they wanted a symbolic record of their actions in the form of standard Logo code to be automatically printed into a text box on the screen (see Figure 10.6). By using the do_history button, students could run commands directly from this text box if they chose – that is if the graphics screen is blank and the history box displays all the commands that made a visual representation of a certain sequence, pressing do_history would allow the user to see the terms of the sequence reconstructed.

We briefly pick out episodes from a case study of a pair of students working with *Mathsticks* to illuminate the principles underpinning our design. Sally and Christine, two 12–13 year-old students were working as a pair at the computer. Their activity began with the opening screen shown in Figure 10.6 and we explained that their task was two-fold: to produce a program that would draw *any* term in the sequence (the seventh and second terms were shown on the screen) and to find a way of working out how many matches would be needed in this general case.

To get started, we suggested they consider the fifth term in the sequence, and asked if they could predict the number of matches for this term. Sally's gestures and verbal descriptions together, indicated – as near as we can tell

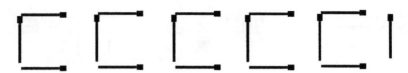

Figure 10.7. Sally's decoding of the sequence.

– that her iconic decoding of the sequence conformed with the incremental pattern shown in Figure 10.7.

Initially Sally's method was not operationalised, and other approaches were initiated by her partner. This meant that by the time the pair turned to Sally's method, Sally had taken on board the idea of constructing a symbolic description of the image she had in mind. The pair started with direct manipulation of the matches in order to obtain the necessary code for Sally's 'C' shape (hmatch match jumpu hmatch). Having done this, they copied the code from the history box, went straight to the procedures page, added the necessary interface jumps and wrote a general procedure:

```
to zig :zag
repeat :zag [hmatch match jumpu hmatch jumpd
umpr]
end
```

The building of a procedure with a variable input appeared completely straightforward; somehow this strange object, zig, was familiar to them because, we can only assume, the meaning of the variable was directly linked to their actions and the way they had visualised the problem. However this program was not quite correct. It left a gap in the final square which was soon debugged and the procedure edited to add an extra match as follows:

```
to zig :zag
repeat :zag [hmatch match jumpu hmatch jumpd
jumpr]
match
end
```

At this point Sally could confidently predict the number of matches for any sequence and could draw the sequence on the screen by calling on zig with an appropriate input. The pair could even write in a text box on the screen a verbal summary of their method as illustrated in Figure 10.8.

A striking facet of the children's work was the extent to which their expressions of relationships, in spoken language, in action and in the final general formulation, were intimately bound into the medium, the situation of which

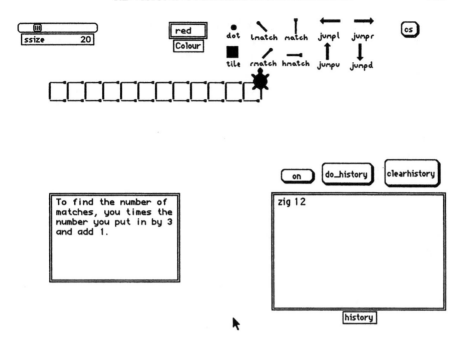

Figure 10.8. Christine and Sally's descriptions of their final generalisation.

they formed a part. It is this intimacy, with its rich set of interconnections between the students, the microworld's activity structures and the mathematical structures of the task, that made progress through the task possible. We designed *Mathsticks* to serve these support functions: to help students to build visual representations of their intended goal by placing computational matches, and in the process to establish a rhythm of action for which symbolic expressions were set up – expressions upon which they could subsequently reflect and act. *Mathsticks* was designed as a domain of situated abstraction, where, following our analysis in Chapter 5, students could sketch their ideas, flag their critical features, use the medium to articulate patterns and build their own templates which could be manipulated and adjusted. Sally and Christine were forging connections between the visual and the symbolic which were globally supportive to the broader aim of writing a general program: they were breathing life into the dormant resources we had placed in the microworld. The dynamic algebra of the programming code became a means for them to think about the relationships in the sequence.

We argue that the microworld's activity structures afforded a theoretical rather than a pragmatic approach; elaboration of structure had a clear pay-off in terms of efficiency. Within *Mathsticks*, the means of expressing actions were firmly soldered to the activity. Virtual matches are easy to connect with real matches, but unlike their real world counterparts, they connect just as easily to both a visual and a dynamic algebra in the system. Of course, just as with real matches, we could not stop students from simply mobilising the *Mathsticks* tools to formulate specific cases in unsystematic ways, or using their results to construct tables of empirical data from which to spot patterns. The point is that within this medium such behaviour would make much less sense; unlike pencil-and-paper drawing, it is efficient to adopt a systematic approach based on the visual structure perceived – in order to draw the sequence and to be able to exploit the `repeat` structure of the programming environment.

2.5. Convivial mathematics

At a recent conference (diSessa, et al., 1995) participants worked on 'The Rugby Problem', taken from a British mathematics textbook (see Figure 10.9). The idea was that people would solve the problem with their preferred choice of tool, try different approaches with different software and finally share all the strategies in a plenary session.

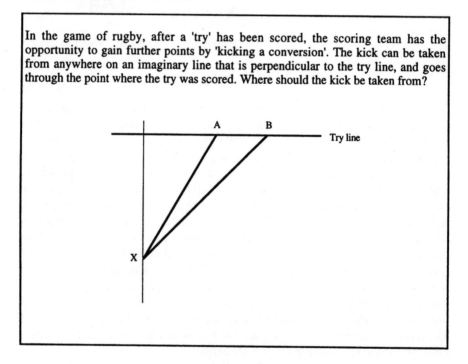

In the game of rugby, after a 'try' has been scored, the scoring team has the opportunity to gain further points by 'kicking a conversion'. The kick can be taken from anywhere on an imaginary line that is perpendicular to the try line, and goes through the point where the try was scored. Where should the kick be taken from?

Figure 10.9. The Rugby Problem.

What was this problem about? The most common view was that it had to do with maximising angles. But this interpretation did not map onto a unique set

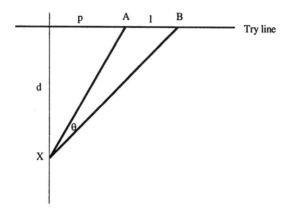

Figure 10.10. A way to parameterise the Rugby problem.

of mathematical techniques. The group who worked with paper and pencil adopted the familiar strategy of drawing diagrams, putting in unknowns, setting up equations and ultimately differentiating to find maximum values. They did not experience the need to use the computer as scaffolding for their work: they knew an approach and were content to stick with it.

In this group, the web of connections was built between the problem and the familiar territory of paper-and-pencil technologies – algebra and calculus. Thus the problem became concretised by being positioned amongst others of a similar nature rather than by its connections to a particular domain of mathematics or to the contextual sense of the setting. The diagrams were simply vehicles for clarifying the variables and the parameters of the problem; in Figure 10.10, these are d, θ, p and l (of course these are not a unique way to parameterise the problem).

The way the problem was to be modelled was instantiated in this process of labelling; it made explicit what was to be ignored and what was to be the focus of attention – the need to find a relationship between θ and d. After this step, the solution, $d^2 = p(p+l)$ for maximum θ, is relatively straightforward – assuming knowledge of and skill at manipulating algebra and calculus tools!

Back at the workshop, most participants eschewed paper and pencil, and went straight to their preferred computer medium to solve the problem. Those who chose to use Cabri activated a web of connections to geometrical objects and their properties. *Their* problem was concretised by *these* meanings: below, we illustrate how far these meanings differed from those of the pencil-and-paper group.

The problem can be modelled in Cabri by constraining A and B to move on the line defining the try line (PAB) and X to move on the perpendicular line, PQ, from which the try can be taken as shown in Figure 10.11. If AX and BX are joined, and X is moved up and down PQ, then the angle AXB

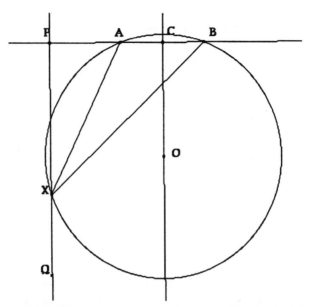

Figure 10.11. The problem is to maximise the angle subtended by the chord AB of the circle which passes through A,B and X, and X lies on PQ.

varies – it is small if PX is large, grows to a maximum and then decreases again. This conjured up the notion that AXB might usefully be considered as the angle at the circumference of a circle subtended by the chord AB; but this prompted the construction of this circle and the consideration of X as the point where it cuts the line PQ (see Figure 10.11).

How could the angle AXB be maximised? Well, AXB is half AOB by the well-known circle theorem. So the question can be restated: At what point is the angle AOB a maximum? Tricky. As O is dragged towards PB, the angle AOB increases but the circle shrinks. There comes a point beyond which the circle does not intersect with PQ at all! So the best that can be done to maximise the angle AOB and ensure that there *is* a point X, is to make PQ the tangent to the circle (see Figure 10.12). Notice how this conjecture emerges from interleaving some familiar geometry facts with data observed through the dynamic interaction with the objects under investigation.

The next step in the solution was to find a way to characterise the position of the centre of the circle, O, in order for PQ to be a tangent. Again, a strategy is suggested by simple geometry. If PQ is a tangent at X, XO is a radius of the circle. But PCOX is a rectangle because of its 90 degree angles; at X from the tangent property, at P from the constraint of the problem and at C from the relationship between the centre of a circle and the mid-point of a chord. Therefore, since opposite sides of a rectangle are equal, PC = OX, and the length of PC is the same as a radius of the circle. But the length of PC is

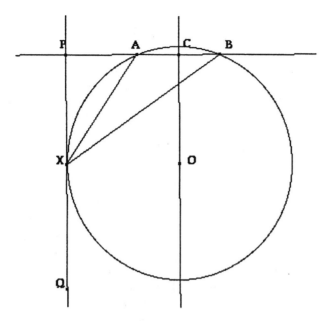

Figure 10.12. O has been dragged until PQ appears to be a tangent to the circle.

given, so the problem is solved by using this length from A or B to pinpoint O.[4] Using software like Cabri renders the solution path strictly geometrical – drawing on geometrical knowledge and exploiting geometrical tools.

Mitch Resnick and Uri Wilensky decided to work on the same problem using their favourite tool, StarLogo. StarLogo comprises thousands of turtles, each of which can control a computational process, with the result that multiple processes can be controlled in parallel. There are many features of StarLogo (these are described in Resnick, 1994) but of principle interest here is that turtles have better senses: they can detect and distinguish other turtles along with certain features of the world over which they move. The StarLogo group, rather than positioning the rugby problem in geometry or calculus, connected it to probability and statistics – a matter of personal preference perhaps, but a preference surely structured by the design of the system they wanted to use. The story of what happened next is taken up by Resnick:

We imagined hundreds of rugby players standing at different points along the line perpendicular to the try line, and we imagined each player kicking hundreds of balls in random directions. To find the point X that maximizes the angle AXB, we simply needed to figure out which of the hundreds of rugby players scored the most conversions (by successfully kicking balls between the goal posts A and B). The reason is clear: If each player is kicking balls in random directions, the player with the largest AXB angle will score the most conversions. This strategy is an example of what is sometimes called a Monte Carlo approach.

It was quite easy to write the StarLogo program to implement this strategy. We used turtles to represent the rugby balls. To start, the program put thousands of turtles/balls with random headings at random positions along the perpendicular line. Then, the program 'kicked' all of these turtles/balls, moving them forward in straight lines. Finally, the program took all of the turtles that successfully went through the goal posts and moved them back to their starting points on the perpendicular line. The point with the most surviving turtles is the point that maximizes the angle AXB.

(Resnick, 1995, pp. 39–40)[5]

What a different view of the problem and one that stunned the workshop with its simplicity! Clearly this solution draws on a very different web of mathematical ideas than the previous two. It is also more closely connected to the original rugby setting, a setting which both the other groups had suppressed the moment they had mathematised the problem by isolating the relevant variables, or locating helpful theorems. Perhaps a more general solution too: Resnick points out 'If extra constraints were added to the problem, such as a wind blowing across the field, or a limitation on the distance a rugby player can kick a ball, it would be quite easy to adjust the StarLogo program to take the new constraints into account. It would be more difficult to adjust traditional geometric analyses' (*ibid.* p. 40).

What can we learn from these three solutions? The first point is that it would clearly seem ridiculous to order the solutions and the competencies displayed in any hierarchy of mathematical sophistication; in each case different connections are made, different meanings are evoked, each coherent and defensible within their own framework.

Let us probe more deeply. In the computer environments a range of situated abstractions are articulated, each encapsulating the way the problem was modelled and how it was solved: in Cabri, the required point is where the circle through AB touches PQ; in StarLogo, it is the one to which most surviving turtles are returned. In contrast, the algebra/calculus solution does not carry along with it any sense of the problem situation; its meanings are derived from connections to a class of problems about optimisation, maxima and minima. In all three solutions, it is evident just to what extent the medium influences the direction of the investigation through the web of connections it evokes and the ways in which ideas are expressed within it.

Why was this activity so successful in the workshop? Because the computer focused our attention on the diversity of knowledge, style and solution. There was also a strong commonality lurking in the subtle domain of feelings and aspirations. Everyone was open to, respected, and was fascinated by the different approaches and epistemological revisions; everyone held the same deeply-rooted affective response to mathematics based on the notions of challenge, empowerment, and enjoyment; everyone wanted to share and compare ideas in an open way either in their construction or in their communication; and everyone adopted a playful approach, tinkering with and experimenting in a culture of building tools to help in the process of conceptualisation. The environment was genuinely collaborative and constructive but also energised by tremendous tenacity tinged with the spice of competition – some worked all night on their solutions determined to show the power of their preferred software! We do not know the extent to which these affective factors determined the diversity of approaches, but we are fairly sure that their absence would at least have dampened them.

A final point. The choice of medium mediates the range of meanings and connections which are likely to structure the interaction, and which are likely to emanate from it. The choice is not arbitrary: it is determined as much by familiarity and expertise as by suitability. All the workshop experts found it virtually impossible to think in a medium other than their favourite. It is simply very difficult to adopt a new perspective, to see through one medium to another. There is a kind of funnelling effect, which is no more evident than with the technology of paper and pencil; and we might find here an explanation for the difficulties which, for example, many mathematicians are currently having in accepting the ways in which the computer is changing mathematics to include an experimental dimension. Yet once the decision is taken to attempt a synthesis, it *can* add another dimension to the problem. Recall the Rugby Problem: the algebraic equation for d, $d^2 = p(p + 1)$ *comes alive* in the geometric configuration. In fact, it represents a simplified expression of the Pythagorean relationship in triangle ACO, where AC is $^1/_2$, CO is d and AO the radius of the circle which has been already established as $(p + ^1/_2)$. The connection confers insight, a commodity in short supply, but surely one which is the ultimate goal of mathematics.

2.6. *Connecting mathematics with practice*[6]

A major multi-national investment bank came to us in May 1994 with a problem. Apparently many of their employees – secretaries, technicians as well as some 'bankers' – had little mathematical appreciation of the models underpinning the financial instruments they were using on a daily basis – not surprisingly given their sophistication. They found difficulty in spotting mistakes and recognising the limitations of the models with the result that the bank was losing money. Mathematics provided the infrastructure of the banking operation but was largely taken for granted or ignored in pursuit of banking objectives. However at times of conflict – and there is nothing so conflictual for a bank as losing money! – the tools became noticed, the implicit shifted to the explicit. The situation described by the bank highlighted the problematic relationship between using a mathematical tool and understanding its essence, a relationship which has previously been studied mainly in the context of elementary mathematics and one which has been a central concern of this book.

Here was a classic situation. How could we assist people to appropriate the mathematics they used on a daily basis? Our first step was to immerse ourselves as far as possible in the practice of banking, to seek to understand the essence of financial mathematics by talking to a wide range of employees, watching them at work, interviewing them *in situ* and reading the literature on the mathematics of finance, trying to make sense of the new language we found there – every field has its own language and banking is certainly no exception. Our struggle was to keep the connection of mathematics with banking, so that meanings from one domain could feed into the other and *vice versa*. This approach was completely novel to our bank-employees. On the one hand, as far as we could tell, financial mathematics courses simply taught the relevant mathematical tools as a system separated from banking practice leaving frustration in their wake. On the other hand, the books we read left most of the mathematical work invisible: it was bypassed with deliberately opaque phrases such as 'it can be shown' or 'this can be proved mathematically'. Such devices inevitably leave the reader in a powerless state – unless they have the resources (intellectual, computational) to fill in the gaps in the text which, as we can now testify, was by no means a trivial task.

Our objective had to be to find ways in which our students could abstract relationships *within* (rather than away from) their working practices – or at least situations which mirrored these practices. We wanted to find a means to respect the classification of instruments and objects used by our students, while bringing to their attention the relationships between them and their underlying structures. In short, the challenge was to find ways of encouraging our students to think about their familiar practices in ways which included the mathematical, rather than by replacing them by 'mathematics'.

There are difficulties associated with this approach. The first was epistemological: how far could we go? To what extent could we provide a new ontology for our students? The second difficulty was pedagogical. If we attempted to get at the underlying mathematical ideas, we would need rigour. But unification and rigour are mathematical priorities, not those of banking practice. Worse still, mathematical rigour is pregnant with meaning, but only for those who are already inducted into its discourse.

Our challenge was to design an environment which afforded rigour with meaning, ways of expressing relationships mathematically without the presupposition that students already knew what we were trying to teach. We decided to adopt the same approach as that which we had tried and tested with children, to open a window for the bank employees onto mathematical ideas through programming.

This was an ideal setting to test out our theoretical and empirical approach, but one that was challenging in its unfamiliarity and high stakes. First, we constructed our own computer models in Logo of a range of simplified finan-

```
to fvs :amount
output :amount * (1 + irate / 100 * days / basis)
end

to pvs :amount
output :amount/(1 + irate / 100 * days / basis)
end
```

10
irate

365
days

365
basis

Figure 10.13. The building blocks: fvs and pvs. Note the values of *irate*, *days* and *basis* are held in "text boxes" (10,365 and 365 respectively).

cial situations. Building models helped us to connect our mathematics to banking, and to sort out what might be the big issues from a mathematical point of view. We trusted that the same process of model building would work in the other direction, connecting for our banker-students their knowledge of banking to mathematics.

Our approach was to view all financial instruments in terms of what *we* viewed as their common mathematical structure, based on the idea of function/program, a perspective which could provide the glue by which different financial instruments could be compared, contrasted and modelled. We planned to immerse students in a programming environment not as a quick way to represent and manipulate data but rather as a means for building representations of relationships symbolically and graphically.[7]

Alongside our intention to make mathematics more visible for our students, we wanted our students to make the practice of banking more visible for us. Thus, our students were co-designers; we wanted to learn (and incorporate into subsequent versions of our materials) how a more systematic, mathematical, view of the banking world could assist in understanding banking practice. We wanted examples from our students which they could work on and explore in the context of their working practices, and from which we could then learn.

Our first task was to distinguish and name a small number of programs as fundamental building blocks for modelling financial instruments. One such building block which epitomises the function/program idea is fvs (**f**uture-**v**alue-**s**imple), which calculates the future value of an amount invested in a simple interest account. We note that since fvs is a function – it has an input and an output – it can be combined with other functions, and in particular has an inverse function, pvs (**p**resent-**v**alue-**s**imple) the essential building block in pricing instruments (see Figure 10.13).

The simplicity of these programs masks a number of interesting programming-mathematical issues. We comment on one: variables can be considered as parameters in the expression of the function and are stored in text boxes; so, for example, irate is the name of a text box whose value is the interest rate;[8] days holds the number of days for the investment; basis specifies the number of days taken to be in a year – which varies across countries between 360 and 365. Since our emphasis lay on functions and graphs as a way of representing a relationship between two quantities, we put considerable stress on the idea that a function has a single input (we were only talking about functions from R to R) and a single output. But we often came upon situations which could be considered as functions of more than one variable: in which case we asked students to choose one as the independent variable and insert the others into text boxes – either by direct manipulation or under program (or command) control. In this way, *all* functions could be displayed graphically, and families of graphs could be produced for different values of the parameters. This approach, which clearly distinguished two mathematical objects – variables as inputs to procedures and parameters held in text boxes – proved important for understanding both mathematics and banking.

Pedagogically, our approach was broadly constructionist (Harel & Papert, 1991). We structured our activities around starting points meaningful to the participants, but which could be seen as jumping-off points for broader investigation and exploration of banking. We tried to incorporate a range of striking examples and intriguing questions. Why should we not expect interest rates to move by full percentage points in 1995 while this was the norm at the turn of the century? What happens to an investment as the frequency of compound interest payments is increased, and why? What is the difference between simple interest and discount instruments?

After the students had constructed and played with fvs and pvs together with their graphs, they were able by a simple interchange of variable and parameter to construct a program, pvs-days, that output the present value (the price) of £100 for a varying number of days to maturity:

```
to pvs-days :days
output 100 / (1 + irate / 100 * :days / basis)
end
```

They could then use the Logo grapher to plot a family of graphs of this function for different interest rates We were interested to see if the students could predict the shape of the curves and wondered if modelling the curves would provide a step to thinking about the relationship between time and quantity: specifically maturity and price. The family of graphs produced is illustrated in Figure 10.14 below, where the horizontal axis is time (0 to 360,000 days[9]) and the vertical axis is price:

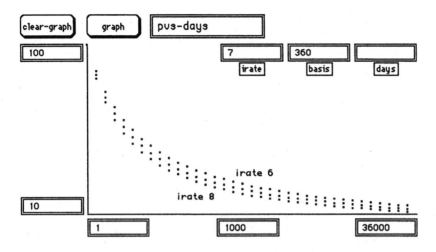

Figure 10.14. A family of price risk graphs.

Each curve shows how price drops with time to maturity. But the family of graphs illuminates much more – most notably that there is a *price risk*; that is, if interest rates increase, there will be a drop in the price of the financial instrument, since, for example, the 7% interest rate curve is always below the 6%. One student, Simon, wanted to go further. He noticed that the gap between the curves was not constant (the vertical distance between them or the change in present value). In other words, Simon made an important discovery: that *price risk varied over time to maturity*. Moreover, he noticed that the size of the gap between the 6% and 7% curves was different than that between the 7% and 8% curves – so price risk varied with prevailing interest rate. We had not noticed this, or if we had, we had not thought it of interest. But Simon had a sense from his practice of these influences on price which tuned his sensitivity in this direction. For example, he was very clear that a fall of one percentage point had a greater impact on securities trading at a lower yield (rate) than on those trading at a higher rate; and that a 100-point fall in rates in a country with a very high interest rate had a less dramatic effect than the same fall in the U.K. Here was the opportunity to look at this phenomenon explicitly rather than tacitly.

Simon wanted to see what happened to price for different times to maturity, when interest rates changed. He wanted to investigate *functions of his functions* by taking a 'cross-section' through his first family of graphs (this description is ours not Simon's) and plotting a new family of differences in price over time for different prevailing interest rates. For him, the curves of price risk had become familiar objects and he had formulated an implicit and powerful meta-theorem: functions reveal structure. This activity spread through the class with some students becoming convinced that there should be a consistent pattern in the price-difference curves, leading them to investigate proportional changes in interest rate (instead of a fixed amount) while others (who knew a little calculus) plotted differentials.[10]

In the real banking world, price risk is considerably more complex than in our model, as interest payments are compounded and the prevailing interest rate is in constant change. Nonetheless, this activity brought price into focus, made it susceptible to mathematical analysis rather than treating it solely as a banking entity. What was so exciting about this work is that it has provided confirmation from a dramatically different context of how to teach and to learn mathematics: how mathematical and banking ideas and their respective motivations can be woven together to produce a powerful and motivating synergy (for us and our students) – a new *BankingMathematics*. The power of the computational modelling approach was that it facilitated this interconnectivity, as students could, to some extent at least, take control of the investigations and shape them according to their banking experiences. Some re-connected with school mathematics, as illustrated by the following comment:

> We designed a program which enabled us to calculate interest rates/days/ future and present values using each different field as a variable. It entailed rearranging the formulae for each calculation and this gave me a better understanding of the formulae themselves and an insight into algebra which I last used at school and thought would never need again!

Others developed new ideas of their practice:

> [The best time was] experimenting with present value curves, both by generating from the machine and by partial differentiation helped me to a greater understanding of the time structure of discount factors. This led me to thinking about how the yield curve translates to a discount factor for each maturity – something which I investigated the following week. And this all stemmed from graphing something which I thought I knew all about already!

There are two final points concerning teaching. We spent a lot of time forgetting about our students as students; we worked on the mathematics together and became excited by this new mathematical field (we recall that this is what happened to some of our teacher-caricatures). We regained an important insight – the extent to which we can understand and teach a new idea is precisely the extent to which we can build a program to express it. Additionally, as we were the teachers, we became even more convinced of the power of epistemological revision to enhance *our own* professional life – it gave us new enthusiasm for what we had thought of as rather simple mathematics, and for the pedagogy we designed around it.

3. RECONNECTING MATHEMATICS AND CULTURE

Our project has been simultaneously to blur the boundaries of mathematics and to broaden the meanings which can be taken to and from it. We have tried, by theory and examples, to challenge the inevitability of one mathematical hierarchy by setting out a range of settings, domains of situated abstraction, where each formulation opens up new possibilities, new definitions and revisions of mathematical practice. If our purpose is to broaden mathematical domains, to construct new kinds of learnable and connected mathematics, then the computer offers an obvious way to set about the task.

From the point of view of school mathematics, the computer can catalyse the opening of new entry points, particularly those from outside school. Our grounds for viewing this optimistically should be clear by now: the tendency to trivialise change, the commodification of knowledge, the 'schoolish epistemology' (diSessa, 1995) which transforms mathematical content into hierarchies and packages, the pressures to disconnect mathematics from all sources of meanings; all these militate against the construction of more learnable mathematics inside schools. But the computer, whether we like it or not, has become, in Papert's memorable phrase, the 'children's machine'; whether we empathise with the nihilistic culture of arcade games or not, children have appropriated them. Even though children neither own nothing of the huge corporate software designers, they *do* own a little of the culture: just enough perhaps to open a chink in the armour which protects access to the portals of the mathematics which underpins it.

The computer points to possibilities. It sets us thinking about ways of reasserting mathematics as a cultural practice, not merely a schoolish activity to achieve technological know-how for the few to pass tests, but to educate in the broadest sense. Perhaps ideas from the culture outside school may be imported, with the computer's help, into school? How might this happen?

We will argue by example, but to do so requires a little historical background. We have seen that Logo has waxed and waned as a fashionable means of learning mathematics in the developed world. But since the collapse of the Soviet Union, a remarkable growth of interest in Logo has occurred among the countries which went to make up its former satellite states. A powerful new version of Logo – *Comenius Logo*[11] – has been written in the Slovak Republic, one which extends the range of mathematical objects which can be manipulated computationally and intellectually (see Blaho, Kalas, & Matusova, 1994). In Bulgaria, a version of Logo which incorporates a euclidean geometry tool kit – *Geomland* – has been developed (we encountered it in Chapter 3) and used as part of a cross-curricular experiment in schools (Sendov & Sendova, 1993). Countries such as Hungary and Poland are exploiting Logo as a means to learn mathematics. There are reasons why this might be so, but they need not concern us here.[12]

What does concern us is that some 12000 copies of Comenius Logo have so far been sold in the Netherlands (Kalas, personal communication, September 1995), and that almost all of these have been to the 'family' market. We are not sufficiently in touch with events in the Netherlands to know why this might be the case, only that the boundary between school knowledge and home knowledge has been blurred, in this case by the advent of a particularly attractive piece of technology (and, perhaps, some rather shrewd marketing on the part of the publisher). The chapters of this book testify that the mere purchase of software hardly provides evidence of a Trojan mouse nibbling away at the school monopoly of mathematical learning. Nevertheless, there are grounds for optimism in the story, in the realisation that some at least are beginning to find alternative uses for their computers than shoot-em-up games, and in the news that schools are apparently beginning to show some interest because of the pressures on them from children and parents.

Our hope is that the meanings of cultures outside school can somehow be connected to meanings constructed in the institutional context. Our theory implies that it is only by activating these webs of connections that we can break down the isolation of school mathematics and transcend its scholastic boundaries. The computer has served as one window on such a possibility, but occasionally we can glimpse others. For example, one of us (CH) co-presented a series of television programmes whose explicit aim was to popularise mathematics. *Fun and Games* was transmitted in England in the prime time slot between 1987 and 1990 with up to 10 million viewers on some occasions. The programmes touched a fascination for mathematics amongst people who had been switched off from it at an early age – judging from the large post-bag and the evaluation questionnaires. Their success showed how a mathematical perspective on a wide range of situations could be introduced through another medium and in everyday language. The lesson of *Fun and Games* is clear – the widespread appeal of mathematics rests in the unique way that it can open up a window on previously unnoticed relationships and structures. We do not claim that *Fun and Games* produced a nation of mathematicians. We do claim that it reconnected mathematics with at least a small piece of the nation's culture.

We recognise that although webbing new mathematical connections is necessary for opening access to mathematics, it is by no means sufficient. We also know it is fraught with difficulty. There are strong pressures to maintain the status quo, and to try to use computers to preserve existing approaches and content, or produce something worse. There is still the widespread belief outside, as well as inside schools that mathematics *is* too difficult, too abstract for general access:

The public wants the concrete; it does not like the abstract, and mathematics is surely one of the most abstract of the human construction.

Boos-Bavnbek and Davis, quoted in Kahane & Choucan (1994, p. 4)

Mathematics learning has become separated from sources of potential meanings; abstraction has come to mean separation and dehumanisation; teachers and students alike have difficulty connecting the meanings of their lived-in cultures with new meanings from outside that culture; and the educational system stands ready to disconnect innovation from any sources of genuine change. What are the sources of this disconnection, and how might they be challenged?

3.1. *The invisibility of mathematical meanings*

The role of mathematics in underpinning social and economic life stretches back to the dawn of the industrial revolution and beyond. Every aspect of modern society is infused with the congealed mathematical labour of mathematicians, computer scientists, engineers and so on. Yet at the same time, this mathematics is increasingly invisible to those who merely share in, rather than construct, the artefacts of the culture – we recall Miss Smilla's apt description of 'reshapers' on the first page of this book. It is mathematics which lies dormant inside the chips of vacuum cleaners, the warheads of missiles, and the graphical displays of news broadcasts. Even the simple exchange of goods and commodities, once relatively amenable to mathematisation, has been overwhelmed by the workings of global markets which are dominated by invisible mathematical forces which are increasingly out of control.[13]

The implications of this state of affairs are far-reaching. At a fundamental level, there has emerged since industrialisation, a situation in which the mathematical needs of the individual and those of society are increasingly at odds. The former requires an appreciation on the part of the many, of the ways mathematics enters into our social, scientific, and technical life as a cultural practice. The latter requires that a few individuals come to wield mathematics as a tool for the reproduction of social and economic life.

This gap is responsible for the ease with which mathematics has come to epitomise all that is not understandable, an excuse for failing to engage with the realities of existence. There is a danger that the symbols of mathematics suffer loss of meaning even as they gain in generality; an obvious source of such loss is the hundreds of special signs and symbols which simultaneously make meanings for the cognoscenti, and obscure them for the rest:

> ... the use of mathematics can become a danger when its only role is to throw dust in your eyes by concealing a lack of thinking concerning the phenomenon under study though a brilliant mathematical development.

> Rapport de conjoncture du Comité National de Recherche Scientifique, 1989: (quoted in Kahane & Chouchan, 1994, p. 9)

Technology can exacerbate this danger by focusing only on those aspects of any situation which can obviously be mathematised:

Mathematical expressions have, however, their special tendencies to pervert thought: the definiteness may be spurious, existing in the equations but not in the phenomena to be described; the brevity may be due to the omission of the more important things, simply because they cannot be mathematised.

Richardson, L.F. (1919) *Mathematical Psychology of War*. (quoted in Hunt & Neunzert, 1994, p. 269)

The disparity between mathematics as a component of culture, and mathematics as a tool for society's use, threatens to turn on mathematics itself, as more and more young people become alienated from what they see as a dehumanised field of study. This disparity feeds strongly back to those whose profession it is to help children learn mathematics; as the cultural forms of mathematics become displaced by pointless curricula and routines – which themselves are ever more divorced from any realities (including mathematical) – these latter become devoid of content, emptied of even the elements of mathematical culture which they may once have held.

The sociologist Basil Bernstein's notions of classification and framing are of some relevance here, as they point to how cultural identities are created and maintained: classification describes the insulating strength of the boundaries between contents, while framing refers to the degree of control teachers and students might exert over the knowledge to be taught. According to this analysis, the weakening of boundaries, which we might refer to positively as the building of new connections, will involve changes in the structure and distribution of power and control: changes in '... what counts as having knowledge, in what counts as a valid realisation of knowledge *and* a change in the organisational context' (Bernstein, 1971, p. 63: emphasis in original).

We should not wonder, therefore, that such change is resisted. We have seen this resistance in action in the foregoing chapters, and it lies at the heart of the challenges we have presented for mathematical education. In order to enter the educational domain, Logo was altered in the image of that domain, fragmented, and compartmentalised into bi-polar opposites of form and content, skills and knowledge. So too were investigations, collaborative work; and so, we predict, will be dynamic geometry, 'groupwork', 'multimedia' and computer algebra systems. In thousands of tiny struggles facing millions of children and teachers, the dilemmas will be the same, shaped by the strong notions students and teachers share of the hierarchies of position and power in institutionalised education.

What is to done? It is all too easy to reach only pessimistic conclusions. Is fundamental resistance to change in education endemic? Do systems necessarily immunise their response systems[14] to defend themselves against change? Are we doomed to a situation in which mathematical meanings remain invisible to all but the élite?

3.2. *Re-visioning learning cultures*

We began with the notion that mathematics pervades our social and physical reality. But it is dead mathematics: the mathematics of the few, designed as much to make invisible to the many the power and limitations of mathematical thought, as it is intended to facilitate the priorities of those who control it.

The technologies that dominate the workplace were mostly designed with this mixture of intentions. On the one hand, to facilitate the performance of routine tasks, to remove the necessity for understanding (including, of course, mathematical understanding) in order to carry them out;[15] on the other, to obscure the realities underlying the routines – whether it be the stock control procedures of the supermarket checkout, or the procedures and consequences of bank transactions.

The coincidence of the economic boom of the nineteen-eighties with the mushrooming of cheap, available and powerful computational technologies, allowed this invasion of the work setting to gather strength. But at the heart of this process there was contained a contradiction, a contradiction which began to work its way to the surface as the economic recession took hold. The assumption that know-how could be embedded in a technology, and manned by a relatively unknowledgeable workforce began to appear a little shaky. At the edges of commerce and industry, especially along those where the financial stakes were highest and the introduction of technology most advanced, some began to doubt the wisdom of putting all their financial eggs into the technological basket, while increasingly marginalising human intellectual resources (see, for example, the analysis by Mitchell, 1989). More than a century ago, Marx famously observed that even the most advanced machines of his day required the application of human labour power to produce wealth. Now, it appears, some are beginning tacitly to recognise the truth of Marx's observation: as the twentieth century draws to a close, it is becoming more evident that intellectual (rather than merely physical) human resources are required to breathe life into technology in order to release its potential. We might interpret the request made to us by the investment bank we encountered earlier in this chapter, as just such a recognition.

And so, ironically, we find in the broader socio-economic arena, the seeds of change which may yet germinate within education. It is here that we propose a final meaning for the notion of re-vision, a making-visible of mathematical structures and relationships which are hidden beneath the surface of our realities. Our analysis indicates that there are ways in which the elements of our mathematised culture can be made visible, a possibility which, as we have seen, resonates with the nascent pressures for visibility in the culture as a whole. It is ironic that the computer is a means – perhaps the only means – to realise this possibility. For the computer allows us to construct settings which synthesise meanings across the cultural divide between application and understanding. Our bankers had no hope of understanding their financial instruments simply by using them, any more than Stevenson's students could understand non-euclidean geometry by watching a demonstration. In both cases, the computer offered students a web of resources which they could connect and construct with what they knew and what they would come to know: it provided an opportunity to make visible new meanings which complemented the old.

Here, at last, is the crucial link between the macro-social and micro-educational forces which might, just might, offer hope for new kinds of mathematical learning. There are plenty of analyses which offer macro-sociological critiques of learning; they, like us, would surely conclude that the neglect of social and cultural sites of meanings, reinforces the notion of mathematics as a dehumanising and élitist activity. But for change to be possible, we must do more than focus exclusively on social and cultural concerns; to do so would imply that individual teachers, researchers and students are powerless in the face of overwhelming social and cultural constraints.

Our notion of a domain of abstraction affords a connection between the mathematical and the cultural. It provides a way to imagine how mathematical structures can be externalised and manipulated within an appropriate symbolic or linguistic framework. Abstractions made in such a setting lack the universality of standard mathematical discourse; but this limitation of epistemology is more than compensated for by the gain in expressiveness. When general relationships can be expressed, they can be explored and become familiar. In the process, the links with knowledge of lived-in-cultures can be maintained, not severed in the quest for ultimate pinnacles of abstraction. The webs of meanings which derive from any domain of abstraction intersect with those derived from other cultural sites. There is a chance to produce a synergy in which both mathematical and cultural practices become more visible.

Grasping the dialectic between individual and social meaning-making affords the possibility of change. It allows us to think of meanings being created within a setting, rather than simply brought into it. We do not have to characterise change agents as thwarted, stymied by habits and routines. Instead, we can see actors in settings making sense of their roles from the interplay of meanings brought to them from diverse practices. It is, we think, feasible to find alternative ways to make mathematical meanings which make sense to and connect with learners. But the implication is clear. If we want to influence the direction in which children learn mathematically, we have to be sensitive to the ways teachers locate new ideas within the panoply of social and cultural relations which make up their professional practice, and remain alert to the multifarious ways that meanings are constructed by the education system and culture – in research, curriculum and schools.

The computer has played a key part in our story. It has offered a window onto the ways in which mathematical learning can become decentred, appreciated as a piece of social and cultural reality rather than an isolated disconnected skill. It has raised the possibility of thinking of mathematical learning environments in which acting-in and coming-to-understand are mutually constitutive. Real change will involve a change in cultures, a reconnection of the functional and cultural roles of mathematics. We believe that the computer can be an agent of reconnection, not a determinant of change in itself.

Our aim is to change school mathematics and perhaps, in time, play some modest role in changing mathematics itself: not in the sense of rejecting the cultural and scientific heritage that underpins mathematical thought, but in the sense of broadening it, findings ways for it to re-enter the intellectual and cultural lives of individuals and groups. In so doing, mathematics could only stand to be enriched by new interplays of meanings from diverse aspects of the social and physical world.

If it did, the world would change too. Not in any mechanistic sense, but in the sense that our knowledge of the world is inevitably mediated by the cultural tools at our disposal. Our knowledge would inevitably be broadened, our scope for understanding – and changing – the world would inevitably increase.

When I was in school I thought that the leaf on a tree was something that was just completely out of the range of mathematical description .... It was just so unusual that none of the mathematical models could apply to it. And now you really look at trees differently. You think of these algorithmic structures that you studied in Logo. And in some sense that's really kind of changed trees.[16]

NOTES

1. Object Logo is a version of Logo which has several unusual and powerful features; it is compiled (so that programs can run relatively quickly) and it allows programmers to adopt the object-oriented paradigm for program construction. This latter feature proved particularly useful in allowing elegant and expressive descriptions of the algorithms underlying the microworld.

2. The work reported in this section was undertaken with Lulu Healy. A full report of the study is to be found in Noss, Healy and Hoyles (in press), on which this extract has been based.

3. *Microworlds Project Builder* is a dialect of Logo which has some object-oriented structures and some aspects of direct manipulation – turtles can be picked up and moved and can speak to each other. It has simple interface features, buttons and sliders, which prove rather transparent ways to help students control their own projects – pressing a button can run a command or a procedure and moving a slider can change a variable.

4. Mathematically, the problem is solved at this point, although to construct the solution in Cabri a geometrical construction or the use of a macro such as *length carry*, is required to 'move' the length PC to a line passing through B.

5. Resnick labels X as K. We have amended his notation for consistency.

6. All the work reported in this section was undertaken in close collaboration with Stefano Pozzi and John Wood. See also Noss and Hoyles (in press). We acknowledge the help of

Al Cuoco through invaluable conversations in the early stages, and Uri Wilensky for his assistance in writing our graphing utility.

7. We used *Microworlds Project Builder* again because it provides a suitable set of primitives so that models could be relatively easily built. Unfortunately it must be admitted that it also (at least on the PC machines we were using) has one almost-fatal flaw: its numerical accuracy is severely limited. We survived this difficulty by a number of 'fixes', none of which was completely satisfactory.

8. In MPB, a text box is like a visible function which simply gives up its value when called. Readers unfamiliar with the idea will not suffer by thinking of irate as the value of a global variable – although this is not actually the case.

9. Time is exaggerated to discern trends more clearly in the graphs.

10. The project briefly outlined above led to the discovery of an apparent maximum price risk (something which nobody in the bank, as far as we can find out, knew before) and its later debugging in terms of banking practice.

11. Variously known by its Dutch publisher as *SuperLogo, HyperLogo*.

12. Part of a possible explanation is that the political possibilities of educational exploration and innovation have been sufficiently delayed in these countries to coincide with computing power and software which can exploit the potential that early versions of Logo were unable to tap.

13. Even, it seems, to those working in the field, as the recent instability caused by the trading of derivatives has revealed.

14. In his book *The Children's Machine*, Papert (1993) uses the metaphor of an immune response system to describe the reaction of schools to change.

15. And, of course, to allow less qualified, cheaper labour power to be employed.

16. Comment by Al Cuoco at the Nato Advanced Research Workshop on Computers for Exploratory Learning, Asilomar California 1993. Quoted in: diSessa, Hoyles, and Noss, (1995) p. 7.

REFERENCES

Abelson, H., & diSessa, A. (1980). *Turtle Geometry: The Computer as a Medium for Exploring Mathematics*. Cambridge, Massachusetts: MIT Press.

Ackermann, E. (1991). From Decontextualised to Situated Knowledge: Revisiting Piaget's Water Level Experiment. In I. Harel & S. Papert (Eds.), *Constructionism* (pp. 269–394). New Jersey: Ablex Publishing Corporation.

Anderson, B. (1986). *Learning with Logo* Microelectronics Education Programme.

Arsac, J., Balacheff, N., & Mante, M. (1992). The Teacher's Role and the Reproducibility of Didactical Situations. *Educational Studies in Mathematics, 23*(1), 5–29.

Assessment of Performance Unit (APU) (1986). *A Review of Monitoring in Mathematics 1978–1982*. London: Department of Education and Science.

Baki, A. (1994). *Breaking with Tradition: a Study of Turkish Student Teachers' Experiences within a Logo-based Mathematical Environment*. Doctoral Dissertation, Institute of Education, University of London.

Balacheff, N. (1986). Cognitive versus situational analysis of problem-solving behaviours. *For the Learning of Mathematics, 6*(3), 10–12.

Balacheff, N. (1993). Artificial Intelligence and Real Teaching. In C. Keitel & K. Ruthven (Eds.), *Learning from Computers: Mathematics Education and Technology* (pp. 131–158). Berlin: Springer-Verlag.

Bauersfeld, H. (1988). Interaction, Construction and Knowledge: Alternative Perspectives for Mathematics Education. In D. A. Grouws, T. J. Cooney, & D. Jones (Eds.), *Effective Mathematics Teaching* (pp. 27–46). Hillsdale, New Jersey: National Council of Teachers of Mathematics, Lawrence Erlbaum Associates.

Baulac, Y., Bellemain, F., & Laborde, J. M. (1994). Cabri Geometry II. Texas Instruments Incorporated.

Becker, H. J. (1982). *Microcomputers in the Classroom: Dreams and Realities*. International Council for Computers in Education.

Becker, H. J. (1987). The Importance of a Methodology that Maximizes Falsifiability: Its Applicability to Research About Logo. *Educational Researcher, June-July*, 11–17.

Behr, M. J., Harel, G., Post, T., & Lesh, R. (1992). Rational Number, Ratio, and Proportion. In D. A. Grouws (Ed.), *Handbook of Research on Mathematics Teaching and Learning: a project of the National Council of Teachers of Mathematics* (pp. 296–333). New York: Macmillan.

Bell (1993). Some Experiments in Diagnostic Teaching. *Educational Studies in Mathematics, 24*, 115–137.

Bell, A. W., Greer, B., Grimison, L., & Mangan, C. (1989). Children's Performance on Multiplicative Word Problems: Elements of a Descriptive Theory. *Journal of Research in Mathematics Education, 20*(5), 434–449.

Bender, P. (1987). Kritik der Logo – Philosophie. *Journal Für Mathematik-Didaktik, 1*(2), 3–103.

Bernstein, B. (1971). On the Classification and Framing of Educational Knowledge. In M. F. D. Young (Ed.), *Knowledge and Control* London: Collier-Macmillan.

Bishop, A. J. (1980). Classroom Conditions for Learning Mathematics. In *Proceedings of the Fourth International Conference for the Psychology of Mathematics Education*, (pp. 338–344). Berkeley, California:

Bishop, A. J. (1988). *Mathematical Enculturation: a cultural perspective on mathematics education*. Dordrecht: Kluwer.

Blaho, A., Kalas, I., & Matusova, M. (1994). Environment for Environments: A New Metaphor for Logo. In J. Wright & D. Benzie (Eds.), *Exploring a New Partnership: Children, Teachers and Technology* (pp. 153 - 166). Amsterdam: North-Holland.

Blaye, A. (1988). *Confrontation Socio-cognitive et Résolution de Problème*. Unpublished Doctoral Dissertation, University of Provence.

Booth, L. R. (1984). *Algebra: Children's Strategies and Errors*. Windsor: NFER-Nelson.

Brock, C., Cappo, M., Arbel, A., Rosin, M., & Dromi, D. (1992). The Geometry Inventor. Scotts Valley, California: Wings for Learning.

Brophy, J. (1986a). Teaching and Learning Mathematics: Where Research Should be Going. *Journal for Research in Mathematics Education, 17*(5), 323–346.

Brophy, J. (1986b). Where are the Data? – A Reply to Confrey. *Journal for Research in Mathematics Education, 17*(5), 361–368.

Brousseau, G. (1984). The Crucial Role of the Didactical Contract in the Analysis and Construction of Situations in Teaching and Learning Mathematics. In H. Steiner (Ed.), *Theory of Mathematics Education* (pp. 110–19). Bielefeld: Institut für Didaktik der Mathematik (IDM), Universität Bielefeld.

Brousseau, G. (1986). Fondements et Méthodes de la Didactique des Mathématiques. *Recherches en Didactique des Mathématiques, 7*(2), 33–115.

Brown, C., Brown, S., Cooney, T. J., & Smith, D. (1982). The Pursuit of Mathematics Teachers' Beliefs. In *Proceedings of the Fourth Annual Meeting of the North American Chapter for the Psychology of Mathematics Education.*, (pp. 203–215).

Bryant, P. (1982). The Role of Conflict and Agreement between Intellectual Strategies in Children's Ideas about Measurement. *British Journal of Psychology*(73), 243–252.

Carpenter, T. P., Fennema, E., & Peterson, P. L. (1986). Cognitively Guided Instruction: An Alternative Paradigm for Research on Teaching. In *Proceedings of the Eighth Annual Meeting of the North American Chapter of the International Group for the Psychology of Mathematics Education.*, (pp. 225–231). East Lansing, Michigan.

Carpenter, T. P., Fennema, E., & Romberg, T. A. (Eds.). (1993). *Rational Numbers: An Integration of Research*. Hillsdale, NJ: Lawrence Erlbaum.

Carraher, T. N., Carraher, D. W., & Schliemann, A. D. (1985). Mathematics in the Streets and in Schools. *British Journal of Developmental Psychology, 3*, 21–29.

Carraher, T. N., Schliemann, A. D., & Carraher, D. W. (1988). Mathematical Concepts in Everyday Life. In G. B. Saxe & R. M. Gearhart (Eds.), *Children's Mathematics* (pp. 71–87). San Francisco: Jossey-Bass.

Cheng, P. W., & Holyoak, K. J. (1985). Pragmatic Reasoning Schemas. *Cognitive Psychology, 17*, 391–416.

Cheng, P. W., Holyoak, K. J., Nesbitt, R. E., & Oliver, L. M. (1986). Pragmatic versus syntactic approaches to training deductive reasoning. *Cognitive Pyschology, 18*, 293–328.

Clarkson, R. (1989). Ratio: Enlargement. In D. C. Johnson (Ed.), *Children's Mathematical Frameworks 8–13: A Study of Classroom Teaching* (pp. 191–217). NFER-Nelson.

Cobb, P. (1990). Multiple Perspectives. In L. P. Steffe & T. Wood (Eds.), *Transforming Children's Mathematics Education: International perspectives* (pp. 200–215). Barcombe, East Sussex: Falmer Press.

Cobb, P., & Bauersfeld, H. (Eds.). (1995). *The Emergence of Mathematical Meaning*. Hillsdale, New Jersey: Lawrence Erlbaum.

Cobb, P., Yackel, E., & Wood, T. (1992). Interaction and Learning in Mathematics Classroom Situations. *Educational Studies in Mathematics, 23*(1), 99–122.

Cockcroft, W. (1994). Can the same Mathematics Program be suitable for all Students?: A Personal View from Mathematics Counts, not forgetting Standards. *Journal of Mathematical Behaviour, 13*, 37–51.

Cockcroft, W. M. (1982). *Mathematics Counts*. London: HMSO.

Confrey, J. (1986). A Critique of Teacher Effectiveness in Research in Mathematics Education. *Journal for Research in Mathematics Education, 17*(5), 347–360.

Confrey, J. (1993). The Role of Technology in Reconceptualizing Functions and Algebra. In J. Becker & B. Pence (Eds.), *North American Chapter of the International Group for the Psychology of Mathematics Education*, (pp. 47–74). California, USA.

Confrey, J. (1995). How Compatible are Radical Constructivism, Sociocultural Approaches, and Social Constructivism? In L. P. Steffe & J. Gale (Eds.), *Constructivism in Education* (pp. 185–225). Hillsdale, New Jersey: Lawrence Erlbaum Associates.

Cuoco, A. (1995). Computational Media to Support the Learning and Use of Functions. In A. diSessa, C. Hoyles, & R. Noss (Eds.), *Computers and Exploratory Learning* (pp. 79–108). Berlin: Springer-Verlag.

Cuoco, A., Goldenberg, P., & Mark, J. (1995). Connecting Geometry with the Rest of Mathematics. In P. A. House & A. F. Coxford (Eds.), *Connecting Mathematics across the Curriculum*. NCTM Yearbook.

Davidov, V. V., Rubtsov, V. V., & Kritsky, A. G. (1989). Psychological Foundations of the Learning Activity Organisation based on a Computer System Usage. In *Children in the Information Age*, (pp. 105–117). Sofia, Bulgaria.

de Corte, E., & Verschaffel, L. (1986). Effects of Computer Experience on Children's Thinking Skills. *Journal of Structural Learning, 9*, 161–174.

Department or Education and Science and the Welsh Office (1991). *Mathematics in the National Curriculum*. London: HMSO.

Department for Education (1995). *Mathematics in the National Curriculum*. London: HMSO.

Desforges, C., & Cockburn, A. (1991). *Understanding the Mathematics Teacher: A Study of Practice in First Schools*. London: Falmer Press.

diSessa, A. (1987). Artificial Worlds and Real Experience. In R. W. Lawler & M. Yazdani (Eds.), *Artificial Intelligence and Education. Volume One: Learning Environments and Tutoring Systems* (pp. 55–78). New Jersey: Ablex Publishing Corporation.

diSessa, A. (1988a). Knowledge in Pieces. In G. Forman & P. Pufall (Eds.), *Constructivism in the Computer Age* (pp. 49–70). Hillsdale, New Jersey: Lawrence Erlbaum Associates.

diSessa, A. (1988b). Social Niches for Future Software. In A. diSessa, M. Gardener, J. Greeno, F. Reif, A. H. Schoenfeld, & E. Stage (Eds.), *Towards a Scientific Practice of Science Education* Hillsdale, New Jersey: Lawrence Erlbaum Associates.

diSessa, A. (1993). Collaborating via Boxer. In P. Georgiadis, G. Gyftodimos, Y. Kotsanis, & C. Kynigos (Eds.), *Logo-like Learning Environments: Reflection and Prospects. Proceedings of the Fourth European Logo Conference*, (pp. 351–357). Athens, Greece: Doukas School.

diSessa, A. (1995). Epistemology and Systems Design. In A. diSessa, C. Hoyles, & R. Noss (Eds.), *Computers for Exploratory Learning* (pp. 15–30). Berlin: Springer-Verlag.

diSessa, A., Hoyles, C., & Noss, R. (Eds.). (1995). *Computers for Exploratory Learning*. Berlin: Springer-Verlag.

Doise, W., Mugny, C., & Perret-Clermont, A. N. (1975). Social Interaction and the Development of Cognitive Operations. *European Journal of Social Psychology*(5), 367–83.

Doise, W., & Mugny, G. (1979). Individual and Collective Conflicts of Centrations in Cognitive Development. *European Journal of Psychology*, *9*, 105–108.

Dowling, P., & Noss, R. (Eds.). (1990). *Mathematics versus the National Curriculum*. Basingstoke: Falmer Press.

Doyle, M. (1993). Logo: Windmills, Writing, and Right Triangles. In P. Georgiadis, G. Gyftodimos, Y. Kotsanis, & C. Kynigos (Eds.), *Logo-like Learning Environments: Reflection and Prospects. Proceedings of the Fourth European Logo Conference*, (pp. 21–32). Athens, Greece: Doukas School.

Dreyfus, T. (1991). On the Status of Visual Reasoning in Mathematics and Mathematics Education. In *Proceedings of the Fifteenth International Conference for the Psychology of Mathematics Education*, (pp. 33–47). Assisi, Italy.

Du Boulay, J. B. H. (1977). Learning Teaching Mathematics. *Mathematics Teaching*, *78*, 52–57.

Dubinsky, E. (1991). Reflective Abstraction in Advanced Mathematical Thinking. In D. Tall (Ed.), *Advanced Mathematical Thinking* (pp. 95–123). Dordrecht: Kluwer Academic Publishers.

Edwards, L.D. (1992). A Logo MicroWorld for Transformation Geometry. In C. Hoyles & R. Noss (Eds.), *Learning Mathematics and Logo* (pp. 127–155). Cambridge, Massachusetts: MIT Press.

Eisenberg, M. (1995). Creating Software Applications for Children: Some thoughts About Design. In A. diSessa, C. Hoyles, & R. Noss (Eds.), *Computers and Exploratory Learning* (pp. 175–196). Berlin: NATO.

Engels, F. (1968). Socialism, Utopian and Scientific. In Lawrence & Wishart (Ed.), *Marx, Engels Selected Works* Moscow: Progress Publishers.

Ernest, P. (1989). The Knowledge, Beliefs and Attitudes of the Mathematics Teacher: A Model. *Journal of Education for Teaching*, *15*(1), 13–33.

Ernest, P. (Ed.). (1994). *Mathematics, Education and Philosophy: An International Perspective*. Brighton: Falmer.

Fasheh, M. (1991). Math in a Social Context: Math with Education as Praxis versus Math within Education as Hegemony. In M. Harris (Ed.), *Schools, mathematics and work* (pp. 57–61). Brighton: Falmer.

Feurzeig, W. (1987). Algebra Slaves and Agents in a Logo-Based Mathematics Curriculum. In R. W. Lawler & M. Yazdani (Eds.), *Artificial Intelligence and*

Education. Volume One: Learning Environments and Tutoring Systems (pp. 27–54). New Jersey: Ablex Publishing Corporation.

Feurzeig, W., Papert, S., Bloom, M., Grant, R., & Solomon, C. (1969). *Programming Languages as a Conceptual Framework for Teaching Mathematics.* Report No. 1889. Bolt, Beranek and Newman.

Fischbein, E., Deri, M., Nello, M., & Marino, M. (1985). The role of implicit models in solving verbal problems in multiplication and division. *Journal for Research in Mathematics Education, 16,* 3–17.

Forman, E. A., & Cazden, C. B. (1985). Exploring Vygotskian Perspectives in Education: The Cognitive Value of Peer Interaction. In J. Wertsch (Ed.), *Culture, Communication and Cognition: Vygotskian Perspectives* (pp. 323–347). Cambridge: Cambridge University Press.

Fullan, M. (1989). Linking Classrooms and School Improvement. *Invited Address to the American Educational Research Association.* San Francisco.

Gerdes, P. (1988). A Widespread Decorative Motif and the Pythagorean Theorem. *For the Learning of Mathematics, 8*(1), 35–60.

Goffree, F. (1985). The Professional Life of Teachers: The Teacher and Curriculum Development. *For the Learning of Mathematics, 5*(2), 26–27.

Goldenberg, E. P., Cuoco, A., & Mark, J. (1993). Connected Geometry. *Tenth International Conference on Technology and Education.* Cambridge: MA.

Good, T., Grouws, D. A., & Ebmeier, H. (1983). *Active Mathematics Teaching.* New York: Longman.

Good, T., Mulryan, C., & McCaslin, M. (1992). Grouping for Instruction in Mathematics: A Call for Programmatic Research on Small Group Processes. In D. Grouws (Ed.), *Handbook of Research on Mathematics Teaching and Learning* (pp. 165–196). New York: MacMillan.

Göranzon, B. (1993). *The Practical Intellect: Computers and Skills.* UNESCO.

Groen, G., & Kieran, C. (1983). In Search of Piagetian Mathematics. In H. Ginsberg (Ed.), *The Development of Mathematical Thinking* (pp. 251–375). New York: Academic Press.

Grouws, D. A. (Ed.). (1992). *Handbook of Research on Mathematics Teaching and Learning: a project of the National Council of Teachers of Mathematics.* New York: Macmillan.

Harel, G., & Confrey, J. (Eds.). (1994). *The Development of Multiplicative Reasoning in the Learning of Mathematics.* Albany: State University of New York Press.

Harel, I., & Papert, S. (Eds.). (1991). *Constructionism.* Norwood, NJ: Ablex Publishing Corporation.

Harris, M. (1987). An Example of Traditional Women's Work as a Mathematics Resource. *For the Learning of Mathematics, 7*(3), 26–29.

Hart, K. M. (Ed.). (1981). *Children's Understanding of Mathematics: 11–16.* London: John Murray.

Hart, K. M. (1984). *Ratio: Children's Strategies and Errors.* Windsor: NFER-Nelson.

Hart, K. M. (1985). *Chelsea Diagnostic Mathematics Tests.* Windsor: NFER-Nelson.

Hatano, G., & Inagaki, K. (1991). Sharing Cognition through Collective Comprehension Activity. In L. B. Resnick, J. M. Levine, & S. D. Teasley (Eds.), *Perspectives on Socially Shared Cognition* (pp. 331–348). Washington DC: American Psychological Society.

Hatfield, L., & Kieren, T. E. (1972). Computer Assisted Problem Solving in School Mathematics. *Journal for Research in Mathematics Education, 3,* 99–112.

Healy, L., Pozzi, S., & Hoyles, C. (1995). Making Sense of Groups, Computers and Mathematics. *Cognition and Instruction, 13*(4), 505–523.

Hillel, J., & Kieran, C. (1987). Schemas Used by 12-Year-Olds in Solving Selected Turtle Geometry Tasks. *Recherches en Didactique des Mathématiques, 8*(1.2), 61–102.

Høeg, P. (1994). *Miss Smilla's Feeling for Snow.* London: Flamingo.

House, E. (1974). *The Politics of Educational Innovation.* Berkeley, California: McCutchan Publishing Corporation.

Howe, C. (Ed.). (1993). Peer Interaction and Knowledge Acquisition. Special edition of *Social Development, 2,* 3.

Howe, J. A. M., O'Shea, T., & Plane, F. (1979). Teaching mathematics through Logo programming: An evaluation study. In R. Lewis & E. Tagg (Eds.), *Computer-Assisted Learning — Scope, Prospects and Limits.* Amsterdam: North-Holland.

Hoyles, C. (1985). What is the Point of Group Discussion in Mathematics? *Educational Studies in Mathematics, 16,* 205–214.

Hoyles, C. (1992). Mathematics Teaching and Mathematics Teachers: A Meta-Case Study. *For the Learning of Mathematics, 12,* 3, 32–44.

Hoyles, C. (1993). Microworlds/Schoolworlds: The Transformation of an Innovation. In W. Dörfler, C. Keitel, & K. Ruthven (Eds.), *Learning from Computers: Mathematics Education and Technology* (pp. 1–17). Berlin: Springer-Verlag.

Hoyles, C. (1995). Exploratory Software, Exploratory Cultures? In A. diSessa, C. Hoyles, & R. Noss (Eds.), *Computers for Exploratory Learning* (pp. 199–219). Berlin: Springer-Verlag.

Hoyles, C., & Healy, L. (1996). *Visualisation, Computers and Learning.* Final Report to the Economic and Social Research Council. Institute of Education, University of London.

Hoyles, C., Healy, L., & Pozzi, S. (1992). Interdependence and Autonomy: Aspects of Groupwork with Computers. In H. Mandel, E. de Corte, S. N. Bennett, & H. F. Friedrich (Eds.), *Learning and Instruction: European Research in an International Context* (pp. 239–257).

Hoyles, C., Healy, L., & Pozzi, S. (1994). Learning Mathematics in Groups with Computers: Reflections of a Research Study. *British Education Research Journal, 20*(4), 465–483.

Hoyles, C., Healy, L., & Sutherland, R. (1991a). Patterns of Discussion between Pupil Pairs in Computer and Non-Computer Environments. *Journal of Computer Assisted Learning, 7,* 210–228.

Hoyles, C., & Noss, R. (Eds.). (1985). *Proceedings of the First Logo and Mathematics Education Conference.* Institute of Education, University of London.

Hoyles, C., & Noss, R. (1987). Children Working in a Structured Logo Environment: From Doing to Understanding. *Recherches en Didactique des Mathématiques, 8*(1.2), 131–174.

Hoyles, C., & Noss, R. (1989). The Computer as a Catalyst in Children's Proportion Strategies. *Journal of Mathematical Behaviour, 8,* 53–75.

Hoyles, C., & Noss, R. (Eds.). (1992). *Learning Mathematics and Logo.* Cambridge, Ma.: MIT Press.

Hoyles, C., Noss, R., & Sutherland, R. (1991b). *The Microworlds Project: 1986–1989* Final Report to the Economic and Social Research Council. Institute of Education, University of London.

Hoyles, C., & Sutherland, R. (1989). *Logo Mathematics in the Classroom*. London: Routledge.

Hughes, M. (1986). *Children and Number*. Oxford: Basil Blackwell.

Hunt, J. C. R., & Neunzert, H. (1994). Mathematics and Industry. In A. Joseph, F. Mignot, F. Murat, B. Prum, & R. Rentschler (Eds.), *First European Congress of Mathematics* (pp. 257–275). Basel: Birkhäuser Verlag.

Illich, I. (1973). *Tools for Conviviality*. London: Calder and Boyars.

Inhelder, B., & Piaget, J. (1958). *The Growth of Logical Thinking from children to adolescents*. New York: Basic.

Isroff, K. (1993). Motivation and CAL in Different Learning Situations. *Paper to World Conference on AI in Education*. University of Edinburgh.

Jackiw, N. (1990). The Geometer's Sketchpad. Berkeley, California: Key Curriculum Press.

Jahnke, H. N. (1983). Technology and Education: the Example of the Computer. *Educational Studies in Mathematics, 14*, 87–100.

Johnson, D. C. (Ed.). (1989). *Children's Mathematical Frameworks 8–13: A Study of Classroom Teaching*. Windsor: NFER-NELSON.

Johnson, D. C. (1991). Algorithmics in School Mathematics: Why, what and how? In I. Wirszup & R. Streit (Eds.), *Developments in School Mathematics Education around the World*, Volume 3. University of Chicago Press.

Johnson, R. (1993). Cultural Studies and Educational Practice. In M. Alvarado, E. Buscombe, & R. Collins (Eds.), *The Screen Education Reader*. Basingstoke: Macmillan Press.

Jones, D. K. (1993). *An Exploratory Study of the Nature and Role of Geometrical Intuition in the Process of Solving Geometrical Problems*. Masters Dissertation, Institute of Education, University of London.

Kahane, J.-P., & Chouchan, M. (1994). Mathematics and the General Public. In A. Joseph, F. Mignot, F. Murat, B. Prum, & R. Rentschler (Eds.), *First European Congress of Mathematics* (pp. 1–22). Basel: Birkhäuser Verlag.

Kaput, J. (1986). Information Technology and Mathematics: Opening New Representational Windows. *Journal of Mathematical Behavior, 5*, 187–207.

Kaput, J. (1992). Technology and Mathematics Education. In D. Grouws (Ed.), *Handbook of Research on Mathematics Teaching and Learning* (pp. 515–556). New York: Macmillan.

Kegan, R. (1982). *The Evolving Self*. Cambridge, MA: Harvard University Press.

Keitel, C. (1986). Cultural Premises and Presuppositions in Psychology of Mathematics Education. *Plenary Lecture to the Tenth International Conference for the Psychology of Mathematics Education*, London.

Keitel, C., Bishop, A., Damerow, P., & Gerdes, P. (Eds.). (1989). *Mathematics, Education and Society*. Paris: UNESCO.

Kilpatrick, J. (1978). Variables and Methodologies in Research on Problem Solving. In L. Hatfield (Ed.), *Mathematical Problem Solving* (pp. 7–20). Columbus, OH: ERIC.

Kilpatrick, J., & Davis, R. B. (1993). Computers and Curriculum Change. In C. Keitel & K. Ruthven (Eds.), *Learning from Computers: Mathematics Education and Technology* (pp. 203–218). Berlin: Springer-Verlag.

Kruger, A. C. (1993). Peer collaboration: Conflict, Cooperation or Both? In C. Howe (Ed.), *Peer Interaction and Knowledge Acquisition* (pp. 165–182). Oxford: Blackwell Publishers.

Kruteskii, V. A. (1976). *The Psychology of Mathematical Abilities in School Children.* Chicago: University of Chicago Press.

Kynigos, C. (1995). Programming as a Means of Expressing and Exploring Ideas: Three Case Studies Situated in a Directive Educational System. In A. diSessa, C. Hoyles & R. Noss (Eds.), *Computers for Exploratory Learning* (pp. 399–420). Berlin: Springer-Verlag.

Laborde, C. (1989). Audacity and Reason: French Research in Mathematics Education. *For the Learning of Mathematics, 9*(3), 31–36.

Laborde, C. (1993a). The Computer as Part of the Learning Environment: The Case of Geometry. In C. Keitel & K. Ruthven (Eds.), *Learning from Computers: Mathematics Education and Technology* (pp. 48–60). Berlin: Springer-Verlag.

Laborde, C. (1993b). Interaction between Social and Mathematical Aspects in the Learning of Mathematics. In *Economic and Social Research Council InTER seminar on Peer Collaboration,* Oxford.

Laborde, C. (1994). Working in Small groups: A Learning Situation. In R. Biehler, R. Scholz, R. Strasser, & B. Winkelmann (Eds.), *Didactics of Mathematics as a Scientific Discipline* (pp. 147–158). Dordrecht: Kluwer.

Laborde, C., & Laborde, J. M. (1995). What About a Learning Environment Where Euclidean Concepts are Manipulated with a Mouse? In A. diSessa, C. Hoyles, & R. Noss (Eds.), *Computers for Exploratory Learning* Berlin: Springer Verlag.

Lancy, D. (1983). *Cross Cultural Studies in Cognition and Mathematics.* New York: Academic Press.

Larsson, S. (1986). Learning from Experience: Teachers' Conceptions of Changes in their Professional Practice. *Journal of Curriculum Studies, 19*(1), 35–43.

Lave, J. (1988). *Cognition in Practice: Mind, Mathematics and Culture in Everyday Life.* Cambridge: Cambridge University Press.

Lawler, R. W. (1985). *Computer Experience and Cognitive Development: A Child's Learning in a Computer Culture.* Chichester: Ellis Horwood.

Lawler, R. W., & Yazdani, M. (Eds.). (1987). *Artificial Intelligence and Education. Volume One: Learning Environments and Tutoring Systems.* New Jersey: Ablex Publishing Corporation.

Light, P. (1983). Social Interaction and Cognitive Development: a Review of Post-Piagetian Research. In S. Meadows (Ed.), *Developing Thinking: Approaches to Children's Cognitive Development.* London: Methuen.

Light, P., & Blaye, A. (1990). Computer-based Learning the Social Dimensions. In H. C. Foot, M. J. Morgan, & R. H. Shute (Eds.), *Children Helping Children.* London: Wiley and Sons.

MacDonald, B. (1991). Critical Introduction: From Innovation to Reform – A Framework for Analysing Change. In J. Rudduck (Ed.), *Innovation and Change: Developing Involvement and Understanding* (pp. 1–13). Milton Keynes: Open University Press.

MacKay, H., & Gillespie, G. (1989). *Extending the Social Shaping of Technology Approach: Ideology and Appropriation.* Department of Behavioural and Communication Studies, Polytechnic of Wales.

Mason, J. (1989). Mathematical Abstraction as the Result of a Delicate Shift of Attention. *For the Learning of Mathematics, 9*(2), 2–8.

Mathews, J. (1989). *Tools for Change: New Technology and the Democratisation of Work.* Sydney: Pluto Press.

Mellin-Olsen, S. (1987). *The Politics of Mathematics Education*. Dordrecht, Holland: Reidel Publishing Company.

Messer, D. J., Joiner, R., Loveridge, N., Light, P., & Littleton, K. (1993). Influences on the Effectiveness of Peer Interaction: Children's Level of Cognitive Development and the Relative Ability of Partners. In C. Howe (Ed.), *Peer Interaction and Knowledge Acquisition* (pp. 279–294). Oxford: Blackwell Publishers.

Mevarech, Z., & Light, P. (Eds.). (1992). *Learning and Instruction, European Research in an International Context*. Oxford: Pergamon Press.

Minsky, M. (1986). *The Society of the Mind*. New York: Simon and Schuster.

Minsky, M., & Papert, S. (1971). *Research at the Laboratory in Vision, Language and Other Problems of Intelligence* Cambridge, Ma: MIT Press.

Mitchell, J. (1989). *Tools of Change: New Technology and the Democratisation of Work*. Marrickville NSW: Pluto.

Moreira, C. (1991). Teachers' Attitudes Towards Mathematics and Mathematics Teaching: Perspectives Across Two Countries. In *Proceedings of the Fifteenth International Conference for the Psychology of Mathematics Education*, 2 (pp. 17–24). Assisi, Italy:

Morgan, C. (1995). *An Analysis of the Discourse of Written Reports of Investigative Work in GCSE Mathematics*. Doctoral Dissertation, Institute of Education, University of London.

Murphy, R. J. L. (1980). Sex Differences in GCE Examination Entry, Statistics and Success Rates. *Educational Studies*, 6(2), 169–178.

Nesher, P., & Tecibal, E. (1975). Verbal Cues as an Interfering Factor in Verbal Problem Solving. *Educational Studies in Mathematics*, 6, 41–51.

Noddings, N. (1990). Constructivism in Mathematics Education. In R. B. Davis, C. A. Maher, & N. Noddings (Eds.), *Constructivist Views on the Teaching and Learning of Mathematics* (pp. 7–19). National Council of Teachers of Mathematics.

Nolder, R. (1990). Accommodating Curriculum Change in Mathematics Teachers' Dilemmas. In *Proceedings of the Fourteenth International Conference for the Psychology of Mathematics Education*, 1 (pp. 167–174). Mexico.

Noss, R. (1985). *Creating a mathematical environment through Programming: a study of young children learning Logo*. Doctoral Dissertation, Chelsea College, University of London.

Noss, R., Healy, L., & Hoyles, C. (in press). The Construction of Mathematical Meanings: Connecting the Visual with the Symbolic. *Educational Studies in Mathematics*.

Noss, R., & Hoyles, C. (1992). Looking Back and Looking Forward. In C. Hoyles & R. Noss (Eds.), *Learning Mathematics and Logo* (pp. 431–468). Cambridge, Massachusetts: MIT Press.

Noss, R., & Hoyles, C. (in press). The Visibility of Meanings: Modelling the Mathematics of Banking. *International Journal of Computers for Mathematical Learning, 1* (1).

Noss, R., Sutherland, R., & Hoyles, C. (1991). *Teacher Attitudes and Interactions* (Final Report of the Microworlds Project: Volume. 2). Institute of Education, University of London.

Nunes, T., & Bryant, P. (1996). *Children doing Mathematics*. Oxford: Blackwell.

Nunes, T., Schliemann, A. D., & Carraher, D. W. (1993). *Street Mathematics and School Mathematics*. Cambridge: Cambridge University Press.

Olson, J. K. (1985). Changing Our Ideas about Change. *Canadian Journal of Education*, *10*(3), 294–307.

Otte, M. (1994). Mathematical Knowledge and the Problem of Proof. *Educational Studies in Mathematics*, *26*(4), 299–321.

Papert, S. (1975). Teaching Children Thinking. *Journal of Structural Learning*, *4*(3), 219–230.

Papert, S. (1980). *Mindstorms. Children, Computers, and Powerful Ideas*. New York: Basic Books.

Papert, S. (1987). Computer Criticism vs Technocentric Thinking. *Educational Researcher*, *16*(1), 22–30.

Papert, S. (1992). Foreword. In C. Hoyles & R. Noss (Eds.), *Learning Mathematics and Logo* (pp. ix-xi). Cambridge, Massachusetts: MIT Press.

Papert, S. (1993). *The Children's Machine*. New York: Basic books.

Pea, R. D., & Kurland, D. M. (1984a). *Logo Programming and the Development of Planning Skills* (Technical Report No. 16). The Centre for Children and Technology, Bank Street College of Education, New York.

Pea, R. D., Kurland, D. M., & Hawkins, J. (1987). Logo and the Development of Thinking Skills. In R. D. Pea & K. Sheingold (Eds.), *Mirrors of Minds: Patterns of Experience in Educational Computing* Norwood, New Jersey: Ablex Publishing Corporation.

Pea, R. D., & Kurland, M. (1984b). On the Cognitive Effects of Learning Computer Programming. *New Ideas in Psychology*, *2*(2), 137–168.

Perret-Clermont, A. N. (1980). *Social Interaction and Cognitive Development in Children*. London: Academic Press.

Phelps, E., & Damon, W. (1989). Problem Solving with Equals: Peer Collaboration as a Context for Learning Mathematics and Spatial Concepts. *Journal of Educational Psychology*, *81*, 639–646.

Piaget, J. (1970). *Genetic Epistemology*. New York: Norton and Norton.

Piaget, J. (1971). *The Psychology of Intelligence*. London: Routledge and Kegan Paul.

Piaget, J., & Inhelder, B. (1968). *The Pychology of the Child*. London: Routledge and Kegan Paul.

Pimm, D. (1987). *Speaking Mathematically: Communication in Mathematics Classrooms*. London: Routledge & Kegan Paul.

Pirie, S., & Schwarzenberger, R. (1988). Mathematical Discussion and Mathematical Understanding. *Educational Studies in Mathematics*, *19*(4), 459–470.

Pirie, S. E. B. (1991). Peer Discussion in the Context of Mathematical Problem Solving. In K. Durkin & B. Shire (Eds.), *Language in Mathematical Education: Research and Practice* (pp. 143–161). Milton Keynes: Open University Press.

PME (1979). Proceedings of the Third International Conference for the Psychology of Mathematics Education. Coventry, England.

PME (1987). Proceedings of the Eleventh International Conference for the Psychology of Mathematics Education. Montreal University.

Polyani, M. (1968). Logic and Psychology. *American Psychologist*, *23*, 27–43.

Pozzi, S., Healy, L., & Hoyles, C. (1993). Learning and Interaction in Groups with Computers: when do Ability and Gender Matter? In C. Howe (Ed.), *Peer Interaction and Knowledge Acquisition* (pp. 222–241). Oxford: Blackwell Publishers.

Resnick, L. (1991). Shared Cognition: Thinking as a Social Practice. In L. Resnick, J. Levine, & S. Teasley (Eds.), *Perspectives on Socially Shared Cognition* (pp. 1–20). American Psychological Association.

Resnick, M. (1994). *Turtles, Termites, and Traffic Jams: Explorations in Massively Parallel Microworlds*. Cambridge, Ma.: MIT Press.

Resnick, M. (1995). New Paradigms for Computing, New Paradigms for Thinking. In A. diSessa, C. Hoyles, & R. Noss (Eds.), *Computers for Exploratory Learning* Berlin: Springer-Verlag.

Restivo, S. (1992). *Mathematics in Society and History*. Dordrecht: Kluwer Academic Publishers.

Robert, A., & Schwarzenberger, R. (1991). Research in Teaching and Learning Mathematics at an Advanced Level. In D. Tall (Ed.), *Advanced Mathematical Thinking* (pp. 127–139). Dordrecht: Kluwer.

Robins, K., & Webster, F. (1989). *The Technical Fix: Education, Computers and Industry*. Basingstoke: Macmillan.

Romberg, T. A. (1988). Can Teachers be Professionals? In D. A. Grouws & T. J. Cooney (Eds.), *Perspectives on Research on Effective Mathematics Teaching* (pp. 224–244). Lawrence Erlbaum Associates, National Council of Teachers of Mathematics.

Roschelle, J. (1992). Learning by Collaboration: Convergent Conceptual Change. *The Journal of the Learning Sciences*, 2(3), 235–276.

Roszak, T. (1986). *The Cult of Information: The Folklore of Computers and the True Art of Thinking*. Cambridge: Lutterworth Press.

Rudduck, J. (1991). *Innovation and Change: Developing Involvement and Understanding*. Milton Keynes: Open University Press.

Sarason, S. (1982). *The Culture of the School and the Problem of Change* (2nd ed.). Boston: Allyn and Bacon.

Saxe, G. (1991). *Culture and Cognitive Development*. Hillsdale, New Jersey: Lawrence Erlbaum Associates.

Scardemelia, M., & Bereiter, K. (1991). Higher Levels of Agency for Children in Knowledge Building: A Challenge for the Design of New Knowledge Media. *The Journal of the Learning Sciences*, 1(1), 37–68.

Schoenfeld, A. (1992). Learning to Think Mathematically: Problem Solving, Metacognition, and Sense Making in Mathematics. In D. A. Grouws (Ed.), *Handbook of Research on Mathematics Teaching and Learning: a Project of the National Council of Teachers of Mathematics* (pp. 334–370). New York: Macmillan.

Seely-Brown, J. S., Collins, A., & Duguid, P. (1989). Situated Cognition and the Culture of Learning. *Educational Researcher* (January/February), 32–42.

Sendov, B., & Sendova, E. (1993). East or West - GEOMLAND is Best, or Does the Answer Depend on the Angle? In A. diSessa, C. Hoyles, & R. Noss (Eds.), *Computers and Exploratory Learning* (pp. 59–78). Berlin: Springer-Verlag.

Sfard, A. (1991). On the Dual Nature of Mathematical Conceptions: Theoretical Reflections on Processes and Objects as Different Sides of the Same Coin. *Educational Studies in Mathematics*, 22, 1–36.

Sfard, A., & Linchevski, L. (1994). The Gains and the Pitfalls of Reification — the Case of Algebra. *Educational Studies in Mathematics*, 26, 191–228.

Silver, E. A. (1992). Mathematical Thinking and Reasoning For All Students: Moving From Rhetoric to Reality. In D. F. Robitaille, D. H. Wheeler, & C. Kieran (Eds.),

Proceedings of the Seventh International Congress on Mathematical Education, (pp. 311–326). Québec: Les Presses de l'Université Laval.

Simmons, M., & Cope, P. (1993). Angle and Rotation: Effects of Different Types of Feedback on the Quality of Response. *Educational Studies in Mathematics, 24*(2), 163–176.

Skovsmose, O. (1994). *Towards a Philosophy of Critical Mathematics Education.* Dordrecht: Kluwer Academic Publishers.

Smith, F. (1986). Insult to Intelligence: The Bureaucratic Invasion of our Classroom Computing. *Journal of Educational Computing, 3*, 101–105.

Smith, J. P., diSessa, A. A., & Roschelle, J. (1993). Misconceptions Reconceived: A Constructivist Analysis of Knowledge in Transition. *Journal of Learning Sciences, 3*(2), 115–163.

Solomon, Y. (1989). *The Practice of Mathematics.* London: Routledge.

Sosniak, L. A., Ethington, C. A., & Varelas, M. (1991). Teaching Mathematics without a Coherent Point of View: Findings from the IEA Second International Mathematics Study. *Journal of Curriculum Studies, 23*(2), 119–131.

Stacey, K. (1992). Mathematical Problem Solving in Groups: Are Two Heads Better than One? *Journal of Mathematical Behavior, 11*, 261–275.

Statz, J. (1973). *The Development of Computer Programming Concepts and Problem-solving Abilities among Ten-year Olds Learning Logo.* Doctoral Dissertation, Syracuse University.

Stevenson, I. (1995). *Constructing Curvature: The Iterative Design of a Computer-Based Microworld for Non-Euclidean Geometry.* Doctoral Dissertation, Institute of Education, University of London.

Suchman, L. A. (1987). *Plans and Situated Actions: The Problem of Human-machine Communication.* Cambridge: CUP.

Swan, M. (1983). *The Meaning and Use of Decimals.* Calculator based Diagnostic Tests and Teaching Materials. Nottingham: Shell Centre for Mathematical Education.

Tall, D. (Ed.). (1991a). *Advanced Mathematical Thinking.* Dordrecht: Kluwer.

Tall, D. (1991b). The Psychology of Advanced Mathematical Thinking. In D. Tall (Ed.), *Advanced Mathematical Thinking* (pp. 3–21). Dordrecht: Kluwer.

Thompson, A. G. (1984). The Relationship of Teachers' Conceptions of Mathematics and Mathematics Teaching to Instructional Practice. *Educational Studies in Mathematics, 15*(2), 105–127.

Thompson, A. G. (1992). Teachers' Beliefs and Conceptions: A Synthesis of the Research. In D. A. Grouws (Ed.), *Handbook of Research on Mathematics Teaching and Learning* (pp. 127–146). New York: Macmillan.

Thompson, P. (1994). The Development of the Concept of Speed and its Relationship to Concepts of Rate. In G. Harel & J. Confrey (Eds.), *The Development of Multiplicative Reasoning in the Learning of Mathematics* (pp. 181–236). Albany, NY: State University of New York Press.

Tournaire, F., & Pulos, S. (1985). Proportional Reasoning: A Review of the Literature. *Educational Studies in Mathematics, 16*, 181–204.

Turkle, S., & Papert, S. (1990). Epistemological Pluralism: Styles and Voices within the Computer Culture. *Signs: Journal of Women in Culture and Society, 16*(1), 345–377.

Underhill, R. G. (1991). Two Layers of Constructivist Curricular Interaction. In E. Von Glasersfeld (Ed.), *Radical Constructivism in Mathematics Education* (pp. 229–248). Dordrecht: Kluwer Academic Publishers.

Vergnaud, G. (1982). Cognitive and Development Psychology Research on Mathematics Education: Some Theoretical and Methodological Issues. *For the Learning of Mathematics, 3*(2), 31–41.

Vergnaud, G. (1983). Multiplicative Structures. In R. Lesh & M. Landau (Eds.), *Acquisition of Mathematics Concepts and Processes* (pp. 127–174). London, New York: Academic Press.

Vergnaud, G. (1987). About Constructivism. In *Eleventh International Conference for the Psychology of Mathematics Education*, 1 (pp. 43–54). Montréal.

Vergnaud, G. (1990). La Théorie des Champs Conceptuels. *Recherches en Didactique des Mathématiques, 10*(2.3), 133–170.

Volosinov, V. N. (1973). *Marxism and the Philosophy of Language* (Ladislav Matejka I.R. Titunik, Trans.). New York: Seminar Press.

von Glasersfeld, E. (1995). *Radical Constructivism in Mathematics Education*. London: Falmer.

von Glasersfeld, E. (Ed.). (1991). *Radical constructivism in mathematics education*. Dordrecht: Kluwer Academic Publishers.

Vygotsky, L. S. (1962). *Thought and Language*. Cambridge, Massachusetts: MIT Press.

Vygotsky, L. S. (1978). *Mind in Society: The Development of Higher Psychological Processes*. Cambridge, Massachusetts: Harvard University Press.

Walker, R. (1981). On Fiction in Educational Research: And I Don't Mean Cyril Burt. In D. Smetherham (Ed.), *Practising Evaluation* (pp. 147–165). Nafferton Books.

Walkerdine, V. (1988). *The Mastery of Reason: Cognitive Development and the Production of Rationality*. London, New York: Routledge.

Weir, S. (1987). *Cultivating Minds: A Logo Casebook*. London: Harper & Row.

Weizenbaum, J. (1984). *Computer Power and Human Reason: From Judgement to Calculation*. Harmondsworth: Penguin Books.

Wertsch, J. V. (Ed.). (1985). *Culture, Communication and Cognition: Vygotskian Perspectives*. New York, USA: Cambridge University Press.

Weyer, S., & Cannara, A. (1975). *Children Learning Computer Programming. Experiments with Langauges, Curricula and Programmable Devices*. (Technical Report No. 250). University of Stanford.

Wilensky, U. (1991). Abstract Mediations on the Concrete and Concrete Implications for Mathematics Education. In I. Harel & S. Papert (Eds.), *Constructionism* (pp. 193–204). Norwood, NJ: Ablex Publishing Corporation.

Wilensky, U. (1993). *Connected Mathematics: Building Concrete Relationships with Mathematical Knowledge*. Doctoral Dissertation, Media Laboratory: Massachusetts Institute of Technology.

Winograd, T., & Flores, A. (1988). *Understanding Computers and Cognition*. Addison Wesley.

Wolf, A. (1984). *Practical Mathematics at Work: Learning Through YTS* (Research and Development Report No. 21). Sheffield, U.K.: Manpower Services Commission.

Wood, D., Bruner, J. S., & Ross, G. (1979). The Rôle of Tutoring in Problem Solving. *Journal of Child Psychology and Psychiatry, 17*, 89–100.

INDEX

273

Mathematics Education Library

Managing Editor: A.J. Bishop, Melbourne, Australia

1. H. Freudenthal: *Didactical Phenomenology of Mathematical Structures.* 1983
   ISBN 90-277-1535-1; Pb 90-277-2261-7
2. B. Christiansen, A. G. Howson and M. Otte (eds.): *Perspectives on Mathematics Education.* Papers submitted by Members of the Bacomet Group. 1986.   ISBN 90-277-1929-2; Pb 90-277-2118-1
3. A. Treffers: *Three Dimensions.* A Model of Goal and Theory Description in Mathematics Instruction – The Wiskobas Project. 1987 ISBN 90-277-2165-3
4. S. Mellin-Olsen: *The Politics of Mathematics Education.* 1987
   ISBN 90-277-2350-8
5. E. Fischbein: *Intuition in Science and Mathematics.* An Educational Approach. 1987   ISBN 90-277-2506-3
6. A.J. Bishop: *Mathematical Enculturation.* A Cultural Perspective on Mathematics Education. 1988
   ISBN 90-277-2646-9; Pb (1991) 0-7923-1270-8
7. E. von Glasersfeld (ed.): *Radical Constructivism in Mathematics Education.* 1991   ISBN 0-7923-1257-0
8. L. Streefland: *Fractions in Realistic Mathematics Education.* A Paradigm of Developmental Research. 1991   ISBN 0-7923-1282-1
9. H. Freudenthal: *Revisiting Mathematics Education.* China Lectures. 1991
   ISBN 0-7923-1299-6
10. A.J. Bishop, S. Mellin-Olsen and J. van Dormolen (eds.): *Mathematical Knowledge: Its Growth Through Teaching.* 1991   ISBN 0-7923-1344-5
11. D. Tall (ed.): *Advanced Mathematical Thinking.* 1991   ISBN 0-7923-1456-5
12. R. Kapadia and M. Borovcnik (eds.): *Chance Encounters: Probability in Education.* 1991   ISBN 0-7923-1474-3
13. R. Biehler, R.W. Scholz, R. Sträßer and B. Winkelmann (eds.): *Didactics of Mathematics as a Scientific Discipline.* 1994   ISBN 0-7923-2613-X
14. S. Lerman (ed.): *Cultural Perspectives on the Mathematics Classroom.* 1994
    ISBN 0-7923-2931-7
15. O. Skovsmose: *Towards a Philosophy of Critical Mathematics Education.* 1994   ISBN 0-7923-2932-5
16. H. Mansfield, N.A. Pateman and N. Bednarz (eds.): *Mathematics for Tomorrow's Young Children.* International Perspectives on Curriculum. 1996
    ISBN 0-7923-3998-3
17. R. Noss and C. Hoyles: *Windows on Mathematical Meanings.* Learning Cultures and Computers. 1996   ISBN 0-7923-4073-6; Pb 0-7923-4074-4

Mathematics Education Library

Managing Editor: A.J. Bishop, Melbourne, Australia

18. N. Bednarz, C. Kieran and L. Lee (eds.): *Approaches to Algebra.* Perspectives for Research and Teaching. 1996

ISBN 0-7923-4145-7; Pb 0-7923-4168-6

KLUWER ACADEMIC PUBLISHERS – DORDRECHT / BOSTON / LONDON